Security Officer's Training Manual

Robert D. Heitert
American National Security Services

REGENTS/PRENTICE HALL, Englewood Cliffs, New Jersey 07632

Library of Congress Cataloging-in-Publication Data

Heitert, Robert D., (date)
Security officer's training manual / Robert D. Heitert.
p. cm.
Includes index.
ISBN 0-13-799420-6
1. Private security services—United States—Handbooks, manuals, etc. 2. Police, Private—United States—Handbooks, manuals, etc.
I. Title.
HV8291.U6H45 1993 92-6543
363.2'89'0683—dc20 CIP

Editorial/production supervision, interior design,
and page makeup: *June Sanns*
Acquisitions editor: *Robin Baliszewski*
Copy editor: *Jim Tully*
Cover design: *Marianne Frasco*
Prepress buyer: *Ilene Levy*
Manufacturing buyer: *Ed O'Dougherty*
Editorial assistant: *Rose Mary Florio*

Published by REGENTS/PRENTICE-HALL
A Division of Simon & Schuster
Englewood Cliffs, New Jersey 07632

Cover photo credit: Night watchman
at AC Sparkplug factory, Flint, MI.
Photo © Michael Hayman/Photo Researchers, Inc., NYC.

Printed in the United States of America

10 9 8 7 6 5 4 3 2 1

ISBN 0-13-799420-6

PRENTICE-HALL INTERNATIONAL (UK) LIMITED, *London*
PRENTICE-HALL OF AUSTRALIA PTY. LIMITED, *Sydney*
PRENTICE-HALL CANADA INC., *Toronto*
PRENTICE-HALL HISPANOAMERICANA, S.A., *Mexico*
PRENTICE-HALL OF INDIA PRIVATE LIMITED, *New Delhi*
PRENTICE-HALL OF JAPAN, INC., *Tokyo*
SIMON & SCHUSTER ASIA PTE. LTD., *Singapore*
EDITORA PRENTICE-HALL DO BRASIL LTDA., *Rio de Janeiro*

Contents

Preface

This manual is written for the individual who wishes to become a professional security officer. It contains basic information necessary not only to attain this goal but also information helpful to you as a security officer in dealings you may have with any other type of law enforcement officer.

It is important to note from the outset that each state has its own laws. Students should thus research and refer to the laws of their own state in matters where these regulations have a direct effect on their actions. Most state laws, especially those involving criminal codes, are similar except possibly in the penalty phase. However, because the penalty phase does not effect the work of the security officer, this aspect is therefore not of consequence.

Enforcement of the law is accomplished by the united efforts of every type of law enforcement agency in the United States. This cooperative effort makes possible a powerful army in the fight against crime. No one person can do it alone. It is the duty of all law enforcement personnel to enforce the laws of the land, to search out members of society who violate these laws, and to act as a deterrent to those who may be thinking of violating the law.

All law enforcement action is not newsworthy, but it is necessary. There are many lesser enforcement requirements that must be attended to but would be left unattended if it were not for the lesser-known member of the enforcement community, namely the security officer. The job of this individual is to protect private property and enforce the rules and regulations of a "private, paying client." In addition, he or she is sworn to uphold the laws of the individual states and of the nation.

Enforcement is the duty of the commissioned law officer, and protection of private property and personnel of a paying client is the duty of a security officer. Although there is a difference in the duties of these members of the law enforcement community, they all wear the same fraternity pin. Hence, lack of cooperation among any members invites a serious threat to the welfare of society.

The student of the security profession should have a high regard for any agency, whether local, state, or federal, whose duty is to uphold and enforce the laws of the land. Naturally, there are times when rivalry between agencies comes into play, but this rivalry should never be allowed to interfere with the ability of all parties to perform their duties.

There are times in the career of a security officer when he or she will come in contact with, and be required to work with, various law enforcement agencies. For this reason it is important for certain elements of police training to enter into the training of a security officer. These elements do not make you a police officer by any stretch of the imagination. You are never to assume such a thing. You have a function of your own, separate and different in many ways from the function of a police officer. Stay in your own backyard until your assistance is called for. But when the time comes, be ready and willing to jump the fence to aid a fellow officer regardless of the uniform or badge he or she is wearing. As part of this training, you will encounter realistic text problems at the end of each chapter in this book. Answers to each problem are contained in the Instructor's Manual.

Finally, the author wishes to acknowledge the helpful comments of those who reviewed the manuscript prior to publication. These include Edward Ryan, Ph.D., Mansfield University, Mansfield, Pa.; David Kassebaum; Gary W. Thompson, Monroe Community College, Rochester, N.Y.; and Julie A. Gilmere, Juris Doctor, Western Illinois University, Macomb, Ill.

Comments from the Author

SECURITY OFFICER REGULATING AUTHORITY BY STATES

The following information is a guide to the authorizing statutes or authority of each state that apply to security personnel. This information is based on a state level and not a local, county, or municipal level. Many counties and cities have their own licensing and training requirements that must be adhered to. The only way to be certain that you are acting within authorized guidelines is to check with the individual local controlling authority. Take nothing for granted.

There is a growing demand for more control and training of security officers that comes with the increased need for additional personnel. The thinking is, simply, that if individuals are going to be working in a field where their work has them dealing with civil or criminal law, they should, or even must, have some training in these matters.

The gap between the commissioned law enforcement officer, who has received many hours of training, and the security officer will only be narrowed when the security officer, either through jurisdictional training or independent study, becomes more knowledgeable, dependable, and reliable in the field.

STATE	AUTHORIZING AUTHORITY	REGISTRATION REQUIREMENTS	TRAINING REQUIRED
Alabama	None	None	None
Alaska	Art.4, AS18.65.400	Yes	None
Arizona	ARS32,Ch.24 & 26	Yes	None
Arkansas	ARK.ST71-2122-71-2159	Yes	Exam
California	PI/ADJ ACT Ch.11	Yes	Exam
Colorado	None	Yes	None
Connecticut	CH.534 Sec.29153-29161	Yes	None
Delaware	DEL.Code Title 24 Ch.13	Yes	None
Florida	FLA.Stat. Ch.493	Yes	None
Georgia	GA.Code Title 43 Ch.38	Yes	8 hours' classroom instruction
Hawaii	HA.Stat. Ch.463	Yes	None
Idaho	None	None	None

STATE	AUTHORIZING AUTHORITY	REGISTRATION REQUIREMENTS	TRAINING REQUIRED
Illinois	ILL.Stat. Ch.111-2601–2639	Yes	Pending
Indiana	IC25-30-1	Yes	None
Iowa	Ch.80A-State Code	Yes	Exam
Kansas	None	None	None
Kentucky	None	Yes	Pending
Louisiana	None	Yes	None
Maine	MRSA Title 32 Sec. 9412	Yes	None
Maryland	MD Code.Art.56, Sec. 79-92	Yes	None
Massachusetts	Gen.Law Ch.147 Sec. 22-30	Yes	None
Michigan	MI Act.330 1968	None	None
Minnesota	Sec. 326.32–.339	None	None
Mississippi	None	None	None
Missouri	SEC84.340 RSMo.1978	Yes	3 days of training and exam
Montana	Mont.State Cod Ch.60 T-37	Yes	Pending
Nebraska	None	Yes	None
Nevada	Ch.648.140	Yes	None
New Hampshire	Ch.106F	Yes	None
New Jersey	N.J.Stat.4519-8–27	Yes	None
New Mexico	Ch.61-27(ff.)	Yes	None
New York	Art.7 Gen.Bus.Law Sec. 70-89a	Yes	None
North Carolina	Ch.74c,Priv. Prot. Sec. Act	Yes	None
North Dakota	43-30-01/16	Yes	Yes
Ohio	Ch.4749	Yes	None
Oklahoma	None	None	None
Oregon	None	None	None
Pennsylvania	Private Detective Act of 1953	None	None
Rhode Island	None	None	None
South Carolina	Act 387	Yes	4 hours of training
South Dakota	None	None	None
Tennessee	None	None	None
Texas	Art.4413 29(bb)VACS	Yes	None
Utah	Senate Bill 95 Security Licensing & Regulation Act, 1979 Gen.Sec.	Yes	Yes
Vermont	Title 26,Ch.59	Yes	Exam
Virginia	Code of Va.54-729.27	Yes	12 hours of training
Washington	None	Yes	None
West Virginia	Ch.30-18	Yes	Yes
Wisconsin	Sec.440.6	Yes	None
Wyoming	None	None	None

Note: Be alert and be sure to check the licensing requirements. States that indicate no requirements for registration or training may leave the licensing and training requirements to the discretion of local jurisdictions. In order to be certain of your own status, you must inquire of the law enforcement authorities of the jurisdiction in which you wish to work.

Some states have legislation pending that could be passed at any time. Thus it is recommended you use the above information as a guide only.

If you have a particular question about the laws, be they civil or criminal, consult the statutues of your own state. In the great majority of cases you will find a set of these statues, in complete and up-to-date form, in your public library. If your library cannot help you, write to the attorney general of your state.

The criminal statutes of most states are very similar. There may be differences in some of the elements of a crime, or the allowable penalties for a particular crime, but for the most part, the differences will be slight.

More important for the security officer is the checking of state statutes on the use of force, and in search and seizure cases. In most instances, these regulations are very similar, but it does not pay to take a chance. Double-check the statues of your own state to be sure.

Foreword

When I first read the *Security Officer's Training Manual* by Robert D. Heitert as a pre-publication manuscript reviewer, I was totally and pleasantly surprised. The surprise was total, in that the book contained the best elements of both a textbook and a manual; and pleasant, in that it was written in a manner that one rarely encounters in either a textbook or a manual.

The rapidly growing profession of security is obviously a very important one. Unfortunately, while it's receiving greater recognition in college and university curricula, books about security either inform in a way that is of little real interest or else are of interest in a way that doesn't really inform. The former tend to present information in a bureaucratic organizational model format that informs without relating pertinent operational information, while the latter tend to be of interest only in the most limited circumscribed areas of specialized application.

As a text, this book is informationally comprehensive, offering action-oriented problem-solving, supplemented with legal citations and references. As a manual, it presents factually-based information with a hands-on description of the subject matter. Further, the book is written in a direct personal manner utilizing vignette techniques to not only introduce, describe and overview the topics, but to visualize them as well. The reader, therefore, will find in this book both a text with a problem-solving narrative, and a manual with an operational visualization of the broad range of issues facing security officers.

The result is more than just a comprehensive, action-oriented, problem-solving operational text, or a factually-based, legal-referenced, directional manual, but rather, a hands-on description of the subject matter written with a personalized directness that brings the information to the reader as it brings the reader to the information: a simultaneous accomplishment not to be found in other texts or manuals on security.

Edward Scott Ryan, Ph.D.
Professor and Internship Coordinator
Department of Criminal Justice
Mansfield University
Mansfield, PA

Security Officer's Training Manual

CHAPTER ONE

A Security Officer's Place in Law Enforcement

A SHORT HISTORY OF THE BEGINNINGS OF PRIVATE SECURITY

In the early 1800s the system of law enforcement in England was totally inadequate. Because of this situation, large industries that could afford to pay the costs hired their own police forces. They enforced the law and order mandated by company management within the company's premises and among its employees. The railroads were particularly active in hiring their own police force.

Law enforcement was little more than nothing when Sir Robert Peel became England's Home Secretary in 1822. Realizing the need for corrective action, Sir Robert worked tirelessly at reforming the system of criminal law in his country. He realized that many private police forces were operating individually and according to their own rules, which was leading to chaos. He saw the need of, and devoted his efforts to establishing, individual community responsibility for the preservation of law and order.

Sir Robert stressed above all that the City of London, because of its position and importance to the nation, should have its own police force comprised of civilians drawn from the community and paid from public tax monies.

Peel proposed to the House of Commons what was known as the Metropolitan Police Act. Passage of the act resulted in the formation of the London Metropolitan Police. Crime prevention was to be its principal purpose.

A rapid increase in the crime rate did not permit the intended purpose to stand for very long. The limited police force found itself spending more and more time investigating crime and apprehending suspects. Crime prevention became almost nonexistent. It soon became apparent that prevention had to come from private means.

When the first colonists arrived in America they brought with them the only form of protection they knew. They used persons who were appointed as constables, and other civilians who were paid by individuals, to roam the streets at night to defend against criminals. They actually earned the name of "night watchmen."

As time went on, the few town sheriffs and marshals had no chance to do more than keep the peace. They had their hands full with drunks, bank robbers, and cattlemen in from long trail drives.

Someone had to protect the railroads, the stagecoaches, the assay offices, and other facilities and conveyances that attracted money and people. To defend themselves and their customers, entities such as these formed their own police units.

In 1851, Allan Pinkerton formed an agency specifically for the protection of railroads. His company was what may be termed the first professional private security agency to operate on a regular basis in the United States.

The Pinkerton Detective Agency, which originated in Chicago, grew in size and reputation during its early days. Allan Pinkerton had many good ideas that aided in crime prevention. The firm's reputation suffered in later years because of misadventures into antiunion activities and strikebreaking. The company survived mainly because by this time it was being run by Robert A. Pinkerton. Robert Pinkerton saw the folly of his father's ways and completely overhauled the operation of the firm.

Pinkerton Inc. still operates as one of the largest (if not the largest) private security agencies in the United States.

WHY THE NEED FOR PRIVATE AND PUBLIC SECURITY?

The need for the services of both public and private security stems from the fact that the country and all of its various subdivisions have laws, statutes, and ordinances governing the conduct and behavior of their citizens. The "codes of conduct," as they may be called, are the will of the people expressed through their elected representatives. Sadly, there are those among us who are not what may be described as "morally responsible." These people find they are not able to live according to accepted norms of society.

Fortunately, these persons are in a minority. This being the case, it is necessary for the minority to be restrained so the majority can live in peace and harmony. The minority of lawbreakers must be weeded out from among the majority law-abiding citizenry. Once separated, the violators can be taught to correct the errors of their ways. Those who can reform will be able to rejoin society and live in harmony with their fellow citizens. Those who cannot must remain segregated from society.

HOW ARE PUBLIC NEEDS TO BE MET, AND BY WHOM?

The needs of the public are well known. For a number of years crime in the United States has skyrocketed. More and more the police forces have dwindled. Individual and private business protection is becoming less available. The costs of police operations rise each year, yet the dollar buys less and people are smothered by taxes. They feel they cannot bear the burden of additional taxes necessary to keep up with the needs of the police to meet professional standards, training, salary, numbers, and specialized equipment.

With this feeling foremost in the minds of all governing agencies and police authorities, how are the needs of the public to be met, and by whom?

Does the answer lie in the hands of private security?

Training, age limitations, and the degree of authority afforded the police have a great deal to do with the differences between public and private security. When these and other subjects are discussed there is usually a large gap between the two. Until this gap is narrowed, there will continue to be great differences in the scope of work of public and private police forces.

THE MAIN DIFFERENCES BETWEEN PUBLIC AND PRIVATE SECURITY

To understand the two most prevalent means by which citizens legally protect their lives and property, one must know and understand the differences between these two means.

We do not intend to discuss mechanical or electronic devices in this book. Our intent is to write about public and private security only. We are interested in the role humans play in the protection of the life and property of the citizenry.

PUBLIC SECURITY

Public security consists of, and is provided by, sworn, deputized *law enforcement officers. They are paid from tax money collected from the public.* Their powers and authority originate from the legal federal, state, county, or municipal government for whom they are employed. The function of these officers is called "law enforcement."

The term "law enforcement" includes the maintenance of public order, the protection of persons and their property from violation, investigation of crimes, and the apprehension of those alleged to have committed crimes.

PRIVATE SECURITY

Private security consists of the protection of the life and property of those who pay for the service from private funds. The private security officer's power and authority are derived from the legal jurisdiction wherein the officer performs his or her services. These jurisdictions may be federal, state, or municipal.

The duties of the private security officer are protective or defensive in nature. Officers act according to the degree of power given them by the governing authority of the jurisdiction in which they work.

Ensuring the safety of lives and property is the motivating force behind private security.

Security can be described as the feeling of well-being resulting from the protection of persons or property from a broad range of hazards including crime, fire, explosion, accidents, sabotage, and civil disturbance. Such hazards can severely limit or disrupt the security of everyone or anyone.

Private security is primarily concerned with artificial hazards, but when hazards are acts of God, security officers are just as concerned.

The identification of contract versus proprietary private security (explained below) can be made through its clients. These are determined in advance of the receipt of services. These clients are businesses or contractors who contract for security services in exchange for a fee.

A very basic element of private security is its funding. Clients pay fees from private funds to private security firms or individuals for their various types of security services.

The security service provided by private security is accomplished through two distinct delivery systems, *proprietary and contractual.*

Proprietary security is the hiring and use of persons employed by an individual company for the exclusive protection of the company's own property and personnel.

Contractual security involves security services that are provided by a private security company to business clients on a contract basis. Security officers are employees of the agency that hires them.

OPINIONS THAT MUST BE CHANGED REGARDING PRIVATE SECURITY

Although numerous commissions have been formed to investigate the potential and future of private security, a large obstacle needs to be removed before any findings can become a reality. The obstacle in question is the general opinion and characterization of the current private security industry.

The public perception of the private security industry must be changed from an almost completely negative impression to a positive one. The notion that private security forces are made up of illiterate, overage, unemployable individuals must be corrected.

Private security forces must lead their own drive to correct this erroneous assumption. They have the power within their grasp to control their own destiny. Through their efforts, owners of security agencies and working security personnel can make possible the emergence of a needed and highly respected profession.

WHERE TO START THE CHANGE

A good place to begin changing opinions in this area is to provide an understanding of the functions of the law enforcement officer and the security officer, and the differences between them. A St. Louis security firm's motto used on its letterhead tells the story in a few simple words: "Law enforcement is the job of the police, security is ours."

Private security has many meanings to many people. The term "private security" describes individual and organizational measures and efforts that provide protection for persons and property. It also describes business enterprises that provide services and products to achieve this protection.

A generally acceptable and explicit definition of private security is difficult to construct. Private security entails the performance of functions and activities for the private sector of society, and some for the public sector. The development of a good working definition of private security is necessary. It should establish the foundation upon which use and understanding of the term is based. Development of a meaningful definition should be made by each individual at the conclusion of this text.

THE FOUNDATION OF PRIVATE SECURITY—THE LAW OF SELF-DEFENSE

By law, private citizens have the right to defend themselves and their property within, and subject to, certain conditions and limitations. This includes defense against theft, destruction of property, damages, assault, trespass, seizure, and all other civil or criminal acts against person or property. It is this right to defend one's self and property that permits one person to hire another person to perform such services.

> The privilege of self-defense rests upon the necessity of permitting a person who is attacked to take reasonable steps to prevent harm to him or herself. The privilege extends to the use of all reasonable force to prevent any threatened harmful or bodily contact, or any confinement, whether intended or negligent. (*Haeussler* v *De Loretto,* 1952, 109 Cal.App.2d, 363,240P2d 654)
>
> The privilege to act in self-defense arises, not only where there is real danger, but also where there is reasonable belief it exists. The defendant is not liable where he acts under a reasonable but mistaken apprehension, the person advancing toward him intends to attack him (*Pearson* v *Taylor,* La.App.1959, 116 So.2d.833) or that the hand which goes into a pocket is reaching for a gun. (*Landry* v *Hill,* La.App.1957, 94 So.2d 308)

The interest of self-protection, "the first law of nature," is perhaps sufficiently important in the mind of the public to justify the result. The belief must be one which a reasonable person would have entertained under the circumstances. The defendant is not required to behave with unusual courage.

THE PRINCIPLE OF SELF-DEFENSE

The definition of self-defense can be stated as the right to use force in order to prevent the illegal or improper use of force against one's self or one's property. Under this definition, two types of force can be used in self-defense, deadly and nonlethal.

> Deadly force is the force necessary, intended, or likely to result in death or serious bodily harm.
>
> Nonlethal force is force such as not to result in death or serious bodily harm.

There is case law from every state to confirm this right of a private citizen; state statutes also make self-defense every citizen's right. However, recall that the restrictions and limitations of this right depend on conditions and circumstances in existence at the time of application. Thus, it is important to remember that this is not an open invitation to do as you wish. As a general rule, you can only use a like or equal force as is being used against you. *Force must stop when resistance stops.*

> The right of self-defense does not accrue to a person until he has availed all proper means to avoid physical combat. (*Commonwealth* v *Peterson,* 275 Mass. 437, 478)
>
> The right of self-defense arises from necessity, and ends when the necessity ends. (*Commonwealth* v *Hartford,* 346 Mass.482, 490)

The laws and conditions that permit people to defend themselves are the same laws and conditions that should permit private citizens to band together under a legally recognized business, and be paid to furnish the required protection or defense to those individuals who are willing to pay private citizens to protect them and their property.

There are heavy liabilities placed on the shoulders of security business owners concerning the use of individuals for this defense. People employed for this purpose must have the proper background, the proper stability, adequate training, and must meet all individual state, county, or municipal licensing requirements.

Company owners and individual security officers must fully understand the laws of self-defense as set by the statutes of their own state. This is especially true when the matter of stealing money or property is involved.

A good rule to follow is, "There is nothing which can be stolen that is worth exchanging a life for."

TASKS OF THE POLICE

The task of detecting and separating the law-abiding from the nonlaw-abiding citizen falls on the shoulders of the various law enforcement officers. They are the "enforcement arm" just as the term *law enforcement officer* suggests. It is they who must search for and follow the evidence involved in a crime, apprehend the suspect, and furnish the prosecution with the evidence and materials needed for presentation to the court by the state or federal prosecutor.

Police officers are the enforcement arm of our country's legal system. They are the legal and authorized protectors of life and property.

The task of law enforcement officers is, by its nature, a difficult one. Each year their job becomes more onerous. Costs associated with maintaining police forces climb at a steady rate while receipts from taxes remain constant or decline.

The personnel-to-tax dollar ratio is almost always out of balance. It can only be returned to balance by reducing police services and staff. Solving things in this manner may make the financial accounting of the jurisdiction look good, but an additional physical and mental burden is placed on the remaining officers.

As the number of police decreases, protection of the public takes precedence over the protection of property. All felony cases take up more of an officer's time. It is not because all felony crime is on the increase, but rather because there are fewer personnel to handle the felony load.

CONCLUSIONS OF THE NATIONAL CRIME COMMISSION

The Task Force on Police of the President's Commission on Law Enforcement and the Administration of Justice arrived at the following conclusion:

> Proposals to relieve the police of what are essentially social services have also been lacking in their consideration of the relationship of such services to the incident of more serious crimes. Domestic disturbances, for example, often culminate in a serious assault or homicide. The down and out drunk is almost a certain victim of a theft if he is left to lie on the street and has any article of value on him. The streetwalking prostitute may, in one sense, be primarily a social problem, but many streetwalkers engage regularly in arranging the robbery of their patron as a supplement to their income.
>
> It might be desirable for agencies other than the police to provide community services that bear no relationship to crime or potential crime situations. But the failure of such agencies to develop the relationship between the social problems in question, and the incident of crime suggests the police are likely to remain for some time as the only 24-hour-a-day, 7-day-a-week agency spread over an entire city in a way which makes it possible for them to respond quickly to incidents of this kind.

THE TASKS OF SECURITY

Police coverage is being spread thinner and thinner. Business and industrial facilities are becoming more aware and alarmed at the situation. More and more private security is coming to the fore. With greater frequency the security officer is being asked to fill some of the voids left by the depleting ranks of police officers.

The duty and job of a security officer is primarily preventive and defensive in nature. The private security officer is to defend and protect the property and personnel of the client for whom his or her employer has agreed to a service contract. The security officer's authority is enforceable within the confines and boundaries of said client's property.

The security officer is obliged to work in cooperation with the police on any felony matter that occurs within the confines of the client's property.

The security officer's training should include, but not be limited to, arrest, use of force, search and seizure, control procedures, public relations, report writing, testimony, firearms, first aid, fire prevention, patrol, and traffic control.

Various commissions researching police protection coverage versus the increase in crime and the dwindling of tax dollars have suggested the possible use of security officers to handle minor incidents. Thus far, these ideas seem to be only in the discussion stage. Before such an idea could be practical, a great deal of legal research would have to be done, and standards and training of security people would have to be upgraded.

It is not the job of security officers to seek out law violators in general. Their job is the protection of the life and property of individuals and businesses, and to act as a deterrent to criminal or civil acts against these individuals or companies.

At no time is a security officer, or a licensed watchman, to be assumed to have the complete authority of a police officer. Security officers may not enforce the law any more than any other citizen, except when within the confines of a client's property.

Security officers are not police officers. They are not intended to look or act like police officers. They have completely different functions from a police officer.

A security officer, in addition to protecting the life and property of a client, has the duty to enforce the rules and regulations of the client within the confines of the client's property. The officer is also to uphold the laws of the jurisdiction in which he or she is operating.

The security officer may use *reasonable force* to protect the life and welfare of the client, the client's employees, and the client's property.

Note: By "reasonable" is meant absolutely no more than is necessary to accomplish the protection mentioned. The use of force may not continue beyond this time. When resistance ceases, the use of force must cease immediately.

A security officer normally has the power to arrest and detain for the proper authorities anyone committing a felony within the boundaries of a client's jurisdiction.

Outside of a client's jurisdiction a security officer has no more authority than an ordinary citizen and must decide on one's own what actions to take in unusual situations when not on duty at a client's facility.

A very important fact for security officers to be aware of is *from where their authority comes and to whom they are responsible for actions performed in the line of duty.*

A security officer normally carries his or her license *at the pleasure of the licensing authority within whose jurisdiction the officer happens to be working,* be this a police authority or a civilian-headed agency of the government.

The privilege of performing one's function within the boundaries of any jurisdiction may be revoked at any time should the security officer be in violation of any legitimate regulation, statute, or ordinance in force within the jurisdiction in question.

Remember: It is imperative for security officers to be thoroughly familiar with all the current regulations of the jurisdiction wherein they intend to perform their duties. Ignorance of these regulations is never an excuse for any violations.

THE NEED FOR COOPERATION

There is an ever-increasing need for security officers who are trained in this specialized field. The need includes persons who are able to understand the scope of their work.

Realizing the scope of his or her specialty, the security officer should stay within these limits.

By conducting themselves within the scope of their work, security officers will not only be welcomed by police officers but will be encouraged and offered any assistance needed by those same officers.

Whether the provision of security services is aimed at private or public interests is a key distinction between public and private security.

Many security concerns, such as crime prevention and the maintenance of order, are common to both private and public security. The degree of emphasis placed on these common concerns and functions provides a distinguishing characteristic between the two. Private security focuses on the prevention and reduction of crime that affects private property. Public law enforcement is primarily concerned with maintaining order, apprehending criminals, and enforcing laws within a constitutionally and statutorily mandated criminal justice system.

Although in theory the goal of public law enforcement agencies is to prevent and reduce crime through efforts tied to the criminal justice system, in practice most of their resources are spent for response to rather than the prevention of crime.

Even though the provision of protective services to the public or private sectors is the basic difference between public and private security, these sectors are often served by both security forces. Private security firms and individuals provide contract security for governmental agencies, facilities, and installations that are elements of the public sector. Similarly, law enforcement agencies, although functioning primarily in the public sector, have increasingly become involved in crime-prevention planning and programs that enlist the cooperation and assistance of citizens and business enterprises.

CHAPTER 1 PROBLEMS

PROBLEM 1

Security officer James Withers is registering his round at the number 7 clock station. The station is located near the main gate to the parking lot of the Watson Trucking Company. As he prepares to punch the clock, Withers's attention is drawn to the sudden sound of excitement coming from the fast food shop across the street from the terminal. Suddenly, amid all the noise, a man runs from the front door of the store with a bag in his hand. For seconds, no one follows the man or even makes an appearance at the door. There were no screams for help nor were shots fired. What should officer Withers do?

CHAPTER TWO

A Basic Explanation of the Law

WHY THE SECURITY OFFICER MUST KNOW BASIC LAW

Private security personnel must be aware of the increasing number of incidents in which they participate, either directly or indirectly, that tie them to some aspect of the state's civil laws, criminal laws, or judicial procedures. They may even be dealing with federal or constitutional law.

While it is not necessary for the security officer to be an expert in the field of law, it is necessary for the officer to be schooled in and knowledgeable of the fundamentals of criminal and civil law. These basics are necessary in order for the officer to realize and understand what he or she is doing, to be fully aware of the seriousness of one's actions, and to have an idea of what results these actions may bring.

To be able to provide legal protection to a client and the client's property, security officers must understand their authority, from whence it derives, and when and under what circumstances they may use it. To act in behalf of a client without knowing the consequences of one's actions is an open invitation to legal proceedings against the officer, his or her employer, and the employer's client.

SUMMARY OF THE TYPES OF LAW

A basic summary of the types of law that govern us, along with an explanation of the principles of legal proceedings, is a good place to start.

Federal Laws are laws enacted by the Congress of the United States. They are called Federal Statutes.

State Laws are laws enacted by the various state legislatures. They are called State Statutes.

Both federal and state laws are divided into two categories, *civil and criminal.*

Note: State and federal laws can also be created by an interpretation by the courts of an already existing law or constitution. Such laws are called *case law* or *common law.*

The distinction between state and federal law lies in the question of jurisdiction. To determine this distinction, one must answer the following questions. Where does the person or persons involved live? Where did the incident take place? Was a state or federal statute involved?

There are instances in which violations overlap both state and federal statutes. Precedent in such situations is usually determined by which branch—the state or the federal government—has the best evidence or the strongest case.

DIFFERENCES BETWEEN CIVIL AND CRIMINAL LAW

Distinct differences exist between civil and criminal law.

CIVIL LAW

Civil law consists of claims between people for alleged injuries, or wrongs committed by a person, or persons, against another person or persons. Damages in the form of monetary payment are usually sought to justify or compensate for these alleged wrongs. There are also instances in civil law in which injunctive relief is sought by one party to prevent another party from performing a particular act. The usual terminology for this injunctive relief is called a *court order* or a *restraining order.* The participants in a civil trial are:

Plaintiff—This is the party bringing the suit against the defendant.

Attorney for the plaintiff—This is the lawyer who represents the person or persons bringing the suit, or suing the defendant.

Defendant—This is the party against whom the suit is brought by the plaintiff.

Attorney for the defendant—This is the attorney who represents the defendant in the suit bought by the plaintiff.

CRIMINAL LAW

Criminal law has to do with the wrongs committed by a person, or persons, against society in general; this could be the public, the state, or the federal government. These wrongs are called *crimes* and are specifically spelled out by the state or federal statutes, or by some common law.

The participants in a criminal trial are:

Defendant—This is the person (or persons) alleged to have committed a crime and who has hired, or has had an attorney appointed by the court, to defend against the allegations made.

Attorney for the defense—This is the lawyer hired by the defendant, or appointed by the court, to defend the accused against the criminal allegations made against him or her by the state or the government.

Prosecutor—This is an attorney who represents the people of the state, and who presents the state or government's case to the court.

In criminal law it is the "people" of a particular state or the United States government that prosecutes the wrongdoer through the person of the prosecuting, circuit, or U.S. attorney, or one or more of their subordinates. The purpose of the prosecution is to convict the wrongdoer, known as the defendant. Defendants found guilty of the charges brought against them, after a fair trial before a judge and jury of their peers, are liable to a penalty. The penalty could be a fine, imprisonment, or both, or probation. This is done in order to deter said defendant or others from committing such a crime in the future.

When crimes are violations of state statutes, the person bringing the charges is known as the *complainant,* and is usually the victim of the crime, a witness who saw the crime take place, or the state. In such instances the complainant theoretically has the right to drop the charges or complaint against the defendant. If the complainant has a change of mind and refuses to prosecute, the state has no case.

When crimes are violations of federal statutes it is somewhat different. The victim brings the case to the United States attorney, at which time, if accepted by the government through its U.S. attorney, *the federal government becomes the complainant.* At that point, the complaining witness has no control over the withdrawal of charges against the defendant.

PRINCIPAL CRIMES FOR THE SECURITY OFFICER TO KNOW

Every state in the United States has its own statutes regarding crimes committed within its jurisdiction. Although the basic elements constituting a crime are usually the same or very similar in most states, the student is cautioned to investigate thoroughly the statutes of his or her own state concerning an individual crime.

Felony: A crime is a felony if a state criminal code says it is, or if the person found guilty can be sentenced to death or imprisonment in a state or federal prison for a term greater than one year. There are various classes of felonies, or varying "degrees" of felonies, in different states.

Misdemeanor: A crime is a misdemeanor if a state criminal code says it is, or if the person found guilty can be imprisoned for a term of one year or less. There are various classes of misdemeanors, or varying degrees of misdemeanors, in different states.

Murder: Ever since the year 1900, the majority of states have all had statutes classifying murder in more than one "degree." Some states have as many as three degrees, or types, of murder. Murder occurs when a person unlawfully and knowingly, recklessly, or negligently causes the death of another human being. It is the difference in the intent that usually determines the degree, as mentioned above. The usual statutory provision defining first-degree murder requires a "deliberate and premeditated killing" or a killing in the perpetration of, or an attempt to perpetrate any felony, or certain special conditions.

Robbery: Robbery is forcible stealing. The usual elements that are incorporated into the modern statutory definition of robbery are the following:

A. a taking and carrying away
B. of the property of another
C. with intent to steal
D. from the person or from the presence of the victim
E. by the use of force against the person or
F. with the threat of the use of imminent force with the intent to compel the victim to acquiesce in the taking and carrying away of the property.

Forcible stealing is committed when a person steals property, and in the course of the theft uses or threatens the immediate use of physical force upon another person for any of the following reasons:

A. preventing or overcoming resistance to the taking of property, or
B. preventing the recapturing of the property immediately after it was taken, or
C. compelling the owner of such property, or another person, to give the property to the perpetrator.

In the greatest majority of states, robbery is classified into two categories. Most of the above information or description covers robbery in general, or what in most states is known as *robbery in the second degree.* The second-degree robbery is usually upgraded to *first-degree robbery* by one of the following aggravating circumstances on the part of the perpetrator:

A. Causes serious physical injury to any person.
B. Is armed with a deadly weapon. (There is no requirement the weapon be actually used.)
C. Uses or threatens the immediate use of a dangerous instrument against any person. (Unlike the situation where a deadly weapon is involved, the dangerous weapon must actually be used, or its use threatened.)
D. Displays or threatens the use of what appears to be a deadly weapon or a dangerous instrument.

Deadly weapons mean any firearm, loaded or unloaded, or any weapon from which a shot, readily capable of producing death or serious physical injury, may be discharged, or a knife, dagger, crossbow, or bow and arrow. Some states have added blackjacks, brass knuckles, and other such items to the list.

Dangerous instruments are any instruments, articles, or substances that, under the circumstances in which they are used, may readily be capable of causing death or other serious physical injury.

Stealing: A person commits the crime of stealing if he or she appropriates (takes, uses, transfers, conceals, or retains) property or services of another, with the purpose to deprive the other person of the property or services, and does so either

A. without the other's consent, or
B. by means of deceit, or
C. by means of coercion.

Note: Most state statutes now make it clear that stealing is not only accomplished by taking the property, but also by using, transferring, concealing, or retaining possession of the property. The statute can be more clearly applied to cases such as embezzlement, fraud, extortion, failure to return rented property, or failure to pay for services. Regardless of whether the property is taken or appropriated in some other fashion, *the intent must be to permanently deprive the owner of its use.*

Trespass: A person commits the crime of trespass if he or she knowingly enters unlawfully, or knowingly remains unlawfully, in a building or a habitable structure, or upon real property. Trespass is handled many ways in various states. Usually there is trespass in the first degree and trespass in the second degree. Since the differences vary so much, it would be best for students to refer to their own state statutes in order to make the proper determination.

Note: Because a security officer comes in contact with this law quite often, it is highly recommended you be certain of all of the facts and conditions of your own state, city, and or municipality regarding the law of trespass.

Enter unlawfully or remain unlawfully—A person enters unlawfully or remains unlawfully, in or on the premises, when he or she is not licensed or privileged to do so. A person who, regardless of the purpose, enters or remains in or upon premises which are at the time open to the public does so with license and privilege unless he or she defies the lawful order not to enter or remain, personally communicated to the person by the owner of said premises, or by any other authorized person. A license or privilege to enter or remain in a building that is only partly open to the public is not a license or a privilege to enter or to remain in that part of the building not open to the public.

Inhabitable structure—This includes a ship, trailer, sleeping car, airplane, or other vehicles or structures,

A. where any person lives or carries on business or other calling, or
B. where people assemble for purposes of business, government, education, religion, entertainment, or public transportation, or
C. which is used for overnight accommodations of persons. Any such vehicle or structure is inhabitable regardless of whether a person is actually present.

THE BASIC ELEMENTS OF CRIMINAL LIABILITY

For a person to be responsible for actions considered to be criminal by law, two conditions must be present:

1. The person's conduct, actions, or possession of an item *must be voluntary.*
2. The person must have the necessary *culpable mental state,* or intent to commit the crime.

WHAT SECURITY OFFICERS SHOULD BE AWARE OF BEFORE TAKING ACTION

For the most part, security officers will not often come in direct contact with a major felony situation during their work. This is not to say it cannot, or will not, happen. There is always the chance. Should this happen, security personnel must know,

A. What type of situation they are facing.
B. What part they are to play.
C. What they are to do and not do.

D. When and whom to call.
E. Their legal responsibilities.

Security officers are more apt to come in contact with misdemeanors or lesser-infraction offenses; in this case they must know all of the above plus the liabilities attached to their actions in the area of civil law.

Courts are more likely to be lenient with a security officer who becomes unintentionally involved in a felony matter because that person usually has not had the amount of formal training or everyday work experience in felony situations as a police officer. Security officers are not members of organized law enforcement agencies of whom more is expected.

CHAPTER 2 PROBLEMS

PROBLEM 1

James is a concert pianist who is on his way to the opera house to give a performance. As he is about to enter the stage door he is approached by two men. They tell him, in no uncertain terms, he either turns over his wallet to them or he suffers two broken hands.

He immediately turns over his wallet, goes into the theater, and phones the police. They arrive and take his report. How do you think this report should be classified? What kind of crime was committed?

PROBLEM 2

In the spaces below indicate with the use of letters whether each statement falls under the jurisdiction of the federal (F), state (S), or local (L) authority. All of the questions involve various infractions or violations of law.

1. _____ No parking, fire lane.
2. _____ All persons between 6 and 16 must attend school.
3. _____ Whoever enters an FDIC bank with intent to steal money or property shall be fined $5,000 or be given 25 years in prison, or both.
4. _____ In order to be a vendor on a public street, one must have a license.
5. _____ All persons flying on interstate flights are subject to search before entering the aircraft departure area.

PROBLEM 3

Using the terms below, fill in the blanks to complete each statement.

civil	criminal	misdemeanor
evidence	prosecutor	plaintiff
defendant	court	precedent
bailiff	clerk	recorder of deeds
preponderance of evidence	felony	judge
beyond reasonable doubt	attorney	trier of fact

Problem: Jim and Andy have been drinking and get into an argument. Jim makes a nasty remark about Andy's girlfriend, and Andy loses his temper. He hits Jim in the jaw and breaks it. Jim sues Andy for his medical expenses, hospital bills, loss of work, and mental suffering.

1. This is a _______ type of case.
2. Jim is the __________.
3. Andy is the __________.
4. Jim must prove his case by a ____________________. The district attorney files charges of assault and battery against Andy. The possible punishment is a fine or imprisonment of up to 6 months, or both.

5. This is a _______ case.
6. The district attorney is the _____________.
7. Andy is the _____________.

PROBLEM 4

In 1963, the case of Gideon v Wainwright was reviewed by the United States Supreme Court (a Florida case).

Facts:
1. Defendant Gideon was charged with breaking into a poolroom.
2. Gideon was poor and asked to be provided with a defense lawyer at no charge.
3. The state court refused his request.
4. The state argued that court-appointed public defenders were available only for serious felonies that involved the death penalty or sentences of life in prison.

1. In this case, what precedent did the U.S. Supreme Court set?
2. Who must follow the precedent?
3. Who would have had to follow the precedent had the case been decided by a state appeals court?

PROBLEM 5

Consider the following cases and decide whether they should be tried in a state or a federal court. Decide to what court should each be appealed.

1. A state sues a neighboring state for dumping waste in a river that borders the two states.
2. A wife sues her husband for divorce.
3. A person is prosecuted for assaulting a neighbor.
4. Two vehicles collide. One driver sues for hospital bills and auto repairs.
5. A group of parents sue the local school board asking their school be desegregated.
6. A man shoots his next-door neighbor.
7. A man commits a robbery in Missouri and transports a hostage across Illinois into Indiana.

PROBLEM 6

This problem involves differences in actions.

Bill likes to tease and annoy people smaller than himself. One day while eating lunch at a local diner Bill sees John picking out a tune on the juke box. Bill orders John to sing along with the record. When John refuses, Bill hits him and breaks his jaw. As a result of his injury, John must pay hospital and doctor's bills and loses three days of work.

1. What law or laws did Bill break?
2. Who would decide if Bill were to be tried for a crime?
3. If Bill were charged with a crime and sued in a civil action, would both charges be tried in one case?

PROBLEM 7

A number of people, including yourself, are stranded on a deserted island after the crash of an overseas air flight. All of the radio equipment is lost, and you have no emergency radio equipment to take its place.

You know it may be some time before you are rescued. Knowing you have been a security officer for 20 years and have some knowledge of the law, the passengers start asking questions.

1. Is this a situation where we need the law?
   Why?
   If we do, do we use existing laws or make up new ones as rules of conduct?
   Why?
2. Draw up five laws to regulate the conduct of the people on the island.
3. Will written laws be enough to regulate conduct?

True or false:

4. In a jury trial of an alleged embezzlement of funds from a bank, the jury determines the applicable law in the trial and the judge rules on questions of fact.
5. The U.S. Supreme Court is the highest court of the land. In most states (but not all) the state supreme court is the highest court.
   State Supreme Courts are the only courts permitted to ignore the precedents of the U.S. Supreme Court.
6. Laws are made only by legislatures.
7. Federal tax dodgers, when brought to trial by the I.R.S. in the State of Florida, will be tried in the local U.S. District Court in Florida.

PROBLEM 8

Late on Thursday night, Jack Armstrong charms his girlfriend, Jane, into driving him to the all-night Quick Shop so he can get a package of cigarettes. When they arrive, Jack goes into the store while Jane waits in the car. Jack pulls out a gun and demands all of the money from the cash register. The night clerk, a man of about 65, gives Jack the money. The clerk then collapses and dies immediately. Jack takes the money and leaves. Jane doesn't find out about the robbery until she hears it on the radio the next morning.

1. Can Jack be charged with homicide?
2. Can Jane be charged with homicide?

PROBLEM 9

Security Officer Ted Barlow is working for a client who engages in interstate shipment of automobiles. The routes of the carrier cover 12 midwestern and southern states.

One evening, while making his rounds among the over-the-road vehicles parked in the loading yards, Barlow notices a brown paper package on the ground under the front end of the trailer frame of one of the vehicles. The package is bound with what appears to be duct tape. It is apparent the package had been attached to the frame of the trailer since one length of the tape is still clinging to one of the hydraulic lift cylinders of the trailer. When the package dropped, part of the corner broke open and a small amount of a white powdery substance fell to the ground.

What should Officer Barlow do?

PROBLEM 10

Melba Wilson has been an armed guard for eight years, has received training at the local police academy, and has shot "expert" each year during her pistol qualifications. Her personnel record is very good in the opinion of the employers for whom she has worked.

On a Tuesday evening in May, Officer Wilson is walking patrol past the front of a liquor store in the West Side Shopping Center. Glancing in the window of the store, she sees an armed robbery in progress. The robber has the clerk and at least 10 other people standing at bay in various locations throughout the store. It is apparent the robber must have entered through the back door, for he is in the rear of the store and had he entered the front door Wilson would have noticed him.

Wilson, seeing the front door is ajar because of the warm spring temperature, draws her revolver and quietly slips up to the opening of the door. She is excited, but feeling she is a good shot, takes aims and fires. Her shot misses and the robber escapes out the back.

Wilson enters the store to pursue the robber, but sees a person on the floor. Blood is coming from a wound on the side of the person's head.

Wilson's shot had hit a pipe near where the robber had been standing and the bullet ricocheted, striking the patron.

What level of culpable mental conduct is Officer Wilson in according to this description of her actions?

CHAPTER THREE

What Goes into Making a Good Security Officer?

The greatest assets one owns, which will help make that individual good security officer, are one's own God-given attributes. These are the abilities that enable a person to expand in personal stature, performance, and importance. They are the greatest factors in the chance for advancement and in the ability to perform one's duties properly on behalf of one's employer.

For the security officer, the most important of these attributes are alertness, imagination, stability, and attitude. From these come the qualities of courtesy, pride, respect, and confidence. These qualities will lead to success in any profession, not only the profession of security.

The material in this chapter is built upon facts that are not really new to the reader but are things of which everyone is aware, but which every so often must be brought to one's attention. The facts are basic and fundamental and are predicated upon common sense. They are the stones of the foundation upon which a person can build a career in the security profession.

PROFESSIONALISM

Professionalism is the only way to improve public opinion concerning the security industry. Only when professionalism is attained will the business community extend to the security profession the recognition it is due. Without this recognition, security personnel will continue to be looked down on, made fun of, and considered an expensive burden.

It is really not difficult to understand the thinking of the business sector when the security officers they encounter are uneducated, uninformed, untrained, slothful, and slovenly.

The first step in attaining the status of professionalism is the ability of each individual security officer to gain the respect of those with whom one comes into contact. This respect is almost always gained as a result of an officer's actions. Nowhere is the demand for the respect of others more necessary than in the fields of security and law enforcement.

Persons in these fields must live as examples to all who look to them for their protection and care. They must actually lead by example.

Society has a right to know that security officers and law enforcement personnel have the highest regard and respect for the laws they are sworn to uphold. These officers cannot expect others to abide by the laws, rules, and ordinances if they themselves do not show a high degree of compliance. By their expression of belief in, and their strict adherence to law and order, they set a positive example that influences those with whom they come into contact. This makes it easier for others to accept the laws. They are not able to offer excuses rather than compliance.

THE ABSOLUTE NEED FOR AN OFFICER TO BE ABLE TO SIZE UP A SITUATION

The professional security officer either has, or must develop, the ability to size up a situation and analyze it. Officers must then be able to decide calmly on a plan of action and resolve the problem in a routine manner.

Problems of major proportions may require more concentration and effort by the officer, but the solution begins the same as mentioned in the prior instance. The only real difference will be in the conclusion. The officer will analyze the situation, assessing it for size and nature. Once the problem is fully evaluated, the officer will know how much and the type of assistance required. Officers will know where and when to obtain this aid. With these facts in hand, the officer will proceed to bring the problem to a conclusion.

ALERTNESS

The first attribute we mentioned at the opening of this chapter was alertness. This may also be called awareness. By either name, it is a prime requisite in becoming a superior security officer and is absolutely necessary for the officer's own safety and well-being, and for the life or property he or she is paid to protect. Knowledge of the conditions, sights, sounds, smells, or activities around one are critical. Everything and everyone, surroundings and terrain, should become the focus of attention.

Each item plays an important part in the officer's ability to analyze a situation instantly. Each has a direct bearing on the action to be taken. It is from within these conditions where the danger to personnel, property, and oneself lies. To be unaware of these things at the moment of confrontation is to invite disaster.

An alert security officer will know the normal and the usual. By knowing the normal and the usual, it is a simple matter to recognize the abnormal and the unusual.

PUTTING ALERTNESS TO PRACTICE

Immediately upon taking his or her post, a security officer should become familiar with everything within sight and hearing. Officers should become aware of any individuals in the vicinity, their habits, their routines, their behavior, their dress, their approximate ages, their sex, and their mode of transportation, if possible. It is not a matter of suspecting everyone in sight or becoming paranoid. Rather, it is simply good, safe practice. Alertness can reduce the element of surprise and become a wealth of information if something should happen later and a possible suspect is described to the officer. Officers will know immediately whether they have or have not seen the suspect.

Officers should make themselves aware of the physical conditions of the buildings and equipment in the immediate area. Their attention should also be directed to the sounds, colors, and smells that are normal to the area. This is a big help in detecting something abnormal. If any of these conditions are contrary to what one knows to be normal, the officer will instantly be on the alert and will almost automatically go into a "defensive mode" while continuing to investigate and analyze what has happened, or is about to happen. Officers will be on guard, and their senses will be fine-tuned for possible trouble. They will be prepared to take any necessary action.

It makes no difference whether the post is new and unfamiliar to the officer, or whether the officer has manned it many times before. Physical checks should be made and mental notes taken every time one reports to the post for duty. Many things may have taken place, many things may have changed or shifted around, or items added or removed, since one's last duty assignment. No matter how small or insignificant any of the items may be, they should be noted.

Possible new fire or safety hazards should be checked. Fire doors and external doors should be examined. Extinguishers, alarms, and other emergency equipment should be gone over and their operational ability assessed. Post order could have been changed since last working the post. All of these things constitute being alert. Incidents happen fast, often without warning. Little time, if any, is allowed to respond to immediate needs. It is too late to look for something, or to use something that is out of order, once trouble starts. Every second saved by prior mental and visual preparation when first arriving for work may mean the difference between life and death. The life could be the officer's. *These truths cannot be emphasized strongly enough.*

ATTITUDE

Attitude is possibly the most important attribute a security officer can offer a client when the client's business requires considerable dealings with the public. *First impressions are usually lasting impressions. Visitors to a client's place of business may be making their first contact with the client through a security officer on duty.* This initial contact can be crucial for future business.

The officer who is courteous, well-groomed, and knowledgeable of his or her duties can make a valuable contribution to the client and to one's own employer. A pleasant, businesslike attitude and an outward indication of agreeableness and cooperation are assets that can propel a security officer to success and higher positions.

Public relations, human relations, and employee relations become an integral part of the security officer's life while on the job. The officer having a positive work attitude will almost certainly be *courteous.* This courtesy will extend not only to the client and the employer, but also to those who work alongside the officer and to all of the client's employees with whom one comes into contact. This courtesy will continue both off the job and on. Courtesy will become a part of the officer's daily life. Success in one's work will be solid evidence that vulgarity, crudeness, and curtness have no place on the job. Instead, these expressions indicate a lack of self-control, lack of ability to control others, and lack of self-esteem on the part of the person so acting. Such behavior is an attempt to mask the individual's inability to discipline oneself.

STABILITY

Stability is usually indicated in people by their willingness to discipline themselves. In security officers, stability permits them to remove personal feelings and their likes and dislikes from their decision making while in the performance of their duties.

Stability permits security officers to conduct themselves properly, and execute orders promptly, without the need of a supervisor being present. These traits separate an ordinary security officer from an exceptional security officer.

By exemplary conduct, both on and off duty, security officers bring credit to themselves and to their employers. Their behavior advertises the fact they rank among the best and have set high standards for themselves. They give notice they are quite capable of high personal performance.

IMAGINATION

Just as imagination plays a large part in the life of every successful businessperson, so it does in the life of a successful security officer. Imagination provides the officer with a tool to cope with future situations before they happen instead of waiting for an event to take place. It helps by aiding in making quick decisions. Imagination permits the officer to plan ahead for things that may happen during the course of one's tour of duty. It provides insight into how and where one will react should the ball come one's way. Imagination has a way of working itself into anticipation. By being able to anticipate a happening, advance preparations can be formulated to handle any situation.

If, during the course of a baseball game, with runners on first and second base, the shortstop waited until the ball was hit to him before deciding what to do with it, it is doubtful his team would win many games. Imagination and anticipation make doing a job much easier than standing idly by waiting for something to happen.

PRIDE

Pride in accomplishment precedes excellence in performance. It is this pride that is essential for security officers to achieve excellence in their work. Pride does not permit failure. If officers have this pride they will accomplish their goal.

Pride gives people the incentive to do their best in every and any job or task they undertake. People such as these find whatever they do enjoyable and find satisfaction in doing a job well. This is not braggadocio pride. It is simply personal pride in doing something right and well.

Find the security officer who takes pride in his or her work and you will also find an officer who has tact. Pride is the thing that brings out the tact to deal with people to the satisfaction of everyone. Tactful security officers choose not to force their will upon others.

APPEARANCE

Appearance plays an important role in the security profession. When security officers put on their uniform, they send a message for all to see. They advertise that they represent a barrier between the client and those who would harm their client's property or personnel. Therefore, they must look the part of an alert, capable, and willing barrier rather than a broken-down stumbling block. The sight of an unkempt, sloppy security officer sends a negative message. It is quite different from the message sent by well-groomed officers. The message the unkempt officer sends is one of indifference. It is a message that says: "I don't really care about the job I'm being paid to do."

If a security officer's appearance impresses a client, or the client's customer, the impression is that the officer performs his or her duties with the same diligence with which he or she dresses. A neat and clean appearance sends a clear message for all to see. The message is "This officer cares." Sloppiness transmits the opposite message: "I don't care."

Uniforms are intended to attract attention and to set the officer apart from others. The attention attracted can be positive or it can be negative. This simple matter of looking clean, sharp, and alert has the power of attracting new customers. A slovenly appearance has the opposite effect of scaring away customers the security contractor already has.

DEDICATION

Businesses cannot attain status in the community unless their owners and employees pay strict attention to their duties, and do so with dedication and devotion. They must be constantly alert to changing times, and be capable of making the necessary adjustments to compensate for those changes. Companies, large or small, are only as good as their employees who do the work and turn out the end product. In this case the company is the security contractor. It is through the efforts of the employees that money is made. By the dedication and devotion of the employees to their daily tasks, company owners and shareholders are paid. Higher education may provide management with the ability to plan and to engineer strategies; money may make the facilities and equipment available to do the work, but the production of the product, be it manufacturing or service, is that produced by employees.

Security officers must have this same dedication and devotion to their duties, to their employers, and their employers' clients. This is the only formula that can provide a strong, firm foundation to the security profession. Agency owners build their reputations on the quality of performance rendered by their employees. This places an obligation on the owners to be honest and fair in their dealings with their personnel, and to be interested in them as people rather than just someone from whom they can derive income.

THE COMMON-SENSE RULE

There are two fundamental rules for a security officer to follow. They are succinct but mean a great deal in the security business. The first is "common sense" and the second is the Golden Rule.

Everyone should know what the Golden Rule says, but the common-sense rule can be broken down into:

Courtesy earns respect,
Knowledge gets results,
Patience gets cooperation,
And service breeds good will.

Like all businesses, security work has a number of rules and regulations basic and peculiar to the industry. They do not pertain to only one agency, but to the entire industry. The longer you work in the business the more you will find that most procedures and practices of the trade are based on common sense and good judgment. They require no college degree or specialized training to put into practice.

BEHAVIOR WHEN DEALING WITH THE PUBLIC

A critical consideration for security officers is their conduct when they come into contact or have dealings with the public. This is a very touchy area. For some it presents no problem, but for others it becomes one of the greatest obstacles to success.

The critical thing for security officers to remember is that they must be *courteous, kind, patient,* and *respectful* in all dealings with everyone. It simply means people want to be treated as human beings who have intelligence and feelings. To be rude to any person without just cause is wrong and has no place in the security business.

Officers must be *impartial* in all judgments. It is true that there are times in the life of a security officer when this becomes a difficult task. But this requirement is an absolute must. There are times when the public can be exasperating, impatient, unkind, and even cruel. Often people do not give thought to what they say or do to others. They blurt out whatever comes into their minds without considering the hurt they may cause. It is at times like these when the security officer must be able to realize what's happening and overlook the public's failings and bad manners. This is not always easy to do, but it is necessary. In such situations it is best to remember that discretion is the better part of valor.

It almost seems as if service personnel are more subject to this kind of treatment than workers in any other industry. In private industry, workers have a limited number of individuals they must work with and please, but in the service business, the worker usually has the entire population to answer to. This is not an ideal situation, but it is something to which service employees must adjust.

Courtesy and a respectful attitude in the face of rude and improper treatment will go a long way in helping to eliminate the problem. "Kill them with kindness" and revealing to the offender the fact you are not intimidated by his or her actions hurts the offender more than anything. Such individuals become shamed and embarrassed without a single unkind word or action from you.

COMMUNICATIONS

A great number of problems with the public, the employer, the client, and with fellow workers can be eliminated. It simply requires the security officer, or anyone else, to realize a communication problem could exist and to make an effort to determine where and how it can be corrected.

Not every person in the world is a born communicator. Many have to work hard and long to overcome communication problems. However, people cannot do anything about something they do not realize exists. They must determine, possibly with the help of others, if they do not hear what others have to say, or if other people do not hear the same thing they are saying. Everyone must be on the same wavelength if effective communication is to take place.

Much of the problem people have in communicating with each other is the fact both sides want to talk at the same time instead of one talking and one listening. Everyone wants to say what one has to say, but no one wants to listen to what the other party has to say. The better a person is at listening, the better he or she will usually be as a communicator.

To be an effective communicator *one must be able to listen with attention, digest the contents of what is heard, and transmit the message clearly and accurately so what comes out of one end must be the same as what goes into the other end.*

Keep this fact in mind: *You can learn little or nothing when your mouth is open, but you can learn a lot when your ears are open.*

When a security officer wants to communicate with others, especially a client, it is really a simple matter. If people have something to say about a subject which they know about, they should address the client in a straightforward manner, without adding frills to try to impress someone. People should speak in an ordinary tone of voice, use simple English, and come straight to the point.

If the client wishes to speak to the officer, the officer should listen attentively to what the client is saying and speak only when the client indicates he or she is through speaking and now expects an answer from the officer.

This is such an important part of the duties of a security officer that it would be wise to obtain copies of any of the many books written on the subject of verbal communication. You may even want to obtain books on body language, the nonverbal means of communication. A great deal can be learned from a person's tone of voice, facial expression, body movements, and gestures. Reading about these things may surprise you and can help make communication and understanding much easier.

CONFIDENCE

Confidence is an attribute important in the lives of all people, but it has special importance in the life of security officers. This air of confidence tells the public the officer knows his or her job, expects to accomplish it, and is prepared for any eventuality.

Officers who are alert to everything and everyone around them, who take pride in their appearance and work, radiate an aura of confidence for all to see. Without knowing it, they will pass this feeling to fellow workers, employers, employers' clients, and everyone with whom they come into contact.

Confidence is not arrogance, cockiness, or trying to make others believe you are something you are not. Just the opposite! Confidence tells others, by your actions and your attitude, that you know what you are doing, that you are good at doing it, and that you can be relied upon to do it properly.

THE SECURITY OFFICER'S MISSION

The mission of security officers is to protect the property and the employees of their employer's client, within the legal confines of each client's property.

This mission is accomplished in numerous ways. The promotion of safety, of employee-management relations, public relations, order, good will, discipline, respect, and confidence are all means by which officers can do their part. They can see to the enforcement of the laws insofar as their client's interests are concerned. They will enforce the client's rules of conduct and safety, policies and procedures within the confines of the client's property.

HOW CAN THE SECURITY OFFICER RENDER PROTECTION TO THE CLIENT?

Security officers can render protection to the client by

1. being constantly on the alert for irregular or unusual conditions or activities;
2. correcting these irregularities when possible, and by reporting all such conditions to the client;
3. permitting only authorized materials to enter or leave the client's premises;

4. permitting only authorized persons in or out of the client's premises;
5. fully enforcing all of the client's rules and regulations;
6. cheerfully rendering service to all;
7. being alert and always planning against potential harm to the clients; and
8. acquiring the respect and good will of the client and the client's employees.

In the proper performance of their duties to protect both client and client's property, security officers will be on the alert and inspect for the following:

1. Fire and safety hazards
2. Evidence of sabotage
3. Materials waste
4. Theft of client's property or employee's property
5. Incendiary or explosives being carried onto client's property
6. Proper badges, identification, and material releases
7. Carrying or use of intoxicants or drugs on client's property
8. Bringing firearms or dangerous weapons by unauthorized persons onto client's property
9. Defective wiring and unnecessary use of electricity
10. Visibility of classified or important documents or materials
11. Good housekeeping and safety practices
12. Unlocked doors, rooms, gates, or containers
13. Misconduct on client's property
14. Loitering on client's property
15. Disturbances
16. Smoking in prohibited areas

Security officers will have their client's best interest in mind when they enforce against the following:

1. Tampering
2. Safety violations
3. Stealing
4. Sabotage
5. Use of drugs or intoxicants on client's property
6. Unauthorized possession of weapons on client's property
7. Misconduct
8. Horseplay
9. Gambling
10. Abuse of equipment
11. Disturbances
12. Smoking in unauthorized areas
13. Littering and unsanitary conditions
14. Loitering

CHAPTER 3 PROBLEMS

PROBLEM 1

Security Officer Mel Hamilton is making his routine clock rounds in the Federal Metals Sheet Metal Company when he hears what sounds like a board cracking or splintering. The only lumber in the place is that used to hold large sheets of aluminum in place, awaiting transfer or sale. The racks are two-tiered high and hold a lot of weight.

Hamilton directs his flashlight beam over the upper racks and thinks it was all his imagination when he sees a split about 4 inches long extending above a joining bolt in the main retaining beams. What should Officer Hamilton do?

PROBLEM 2

Martha Hess is the receptionist at the exclusive Harreson Wholesale Jewelers Company. She has a prestige job, but in addition it is her responsibility to be on the lookout for suspicious persons acting as customers, or persons casing the place through the plate glass windows. Whenever she sees someone or something suspicious, Martha is to push the button under her desk so security can take over from her.

Martha has been trained in security and knows how floor planning and proper displays can play a big part in deterrence.

Martha has a serious gall bladder attack and is out of the store for eight weeks. When she returns she finds things changed greatly around the front store area, which is her assignment. What should she do?

PROBLEM 3

Security Officer Hamilton, while making a tour of the Press Well Engineering Company plant, detects a strange odor coming from a storage area containing 55-gallon drums of chemicals. The plant has been shut down since 5:00 P.M. and it is now 10:30 P.M.

Hamilton directs the beam of his flashlight into the storage area, and after a moment or so detects a small stream of an unknown fluid leading back into the storage area. He follows the stream of fluid back in among the drums, and after a few moments he locates from where it is coming. He spies a 55-gallon drum that has a large dent in its side and which is leaking fluid. The dent seems as though it was caused by the fork of a tow motor bumping into the drum while being moved.

Press Well Engineering makes plastics and chemicals for industry and is known to use and have on hand at various times stocks of volatile and flammable liquids. What should Officer Hamilton do?

PROBLEM 4

Security Officer James Olsen stands guard on the midnight to 8:00 A.M. shift in the lobby of the Goldman Fur Company. The entrance to the main storage vaults is there. Jim has been stationed here for six years and knows every time the floor-length drapes are cleaned, or the potted plants grow a fraction of an inch. Each time he comes on duty and relieves the person before him, the first thing he does is make a tour of the premises to see whether everything is secure.

On the evening of February 1, when Jim entered, he began to make his tour. Half way through it, he hurried to the phone and called the police.

When the police arrived they checked behind the drapes and found the owner on the floor behind the drapes, dead. The police asked Olsen how he knew the owner's body was there since it wasn't visible without moving the drapes. What do you think Jim's answer was?

CHAPTER FOUR
Types of Services and Their Differences

Private security contractors and their hired personnel are part of the "service industry." Often people and companies in this general category of business do not give a thought to the services they perform. They do not stop to think about the relationship that exists between themselves and those for whom they perform their service.

Security personnel, being in the category of service employees, often have responsibilities placed on them that are not usually required of personnel in other service sectors. Without knowing or realizing what or how important these responsibilities are makes success in the security officer's field of service an impossibility.

A DEFINITION OF SERVICE

Service, as defined in Webster's, is "work done for, and benefit conferred on another." This definition seems adequate on the surface, but closer examination reveals it falls short of our needs. It appears too general and does not provide the explanation we need.

By making some changes in Webster's definition, perhaps a better understanding of the term will result. First, subtract the words "benefit conferred on another" and substitute this with the definition of "benefit." Webster defines "benefit" as "to be useful to another." By making this substitution we have a new definition. Putting it together we have "Service is work done to be useful to another."

To be specific in our approach to the act of performing service, we will single out the "security service."

Having singled out the security service and applying our new definition of "service," we see it defines what should be the aim of every security officer. Actually, it describes the foundation upon which the security service is built. The dedication of a responsible security officer to his or her profession is best described in that part of the definition that reads "to be useful to another." This means that the rendering of a service requires thought and deliberation on the part of the person being useful. There is more to being useful to another person than most people realize. Close examination of what is required of the doer reveals that *emotions come into play in a large way. This includes emotions of the person performing the service and the feelings of the person(s) receiving the service.*

WHAT IS INVOLVED IN THE SERVICE BUSINESS?

In 1979, a researcher named J. K. Knisley was overheard during a conversation to say, "In the service business you are dealing with something that is primarily delivered by people, to people. *Your people are as much of the product, in the customer's mind, as any other attribute of service, and people's performance, day in and day out, fluctuates.*"

This researcher put a finger on the pulse of the service business. At the most basic and elementary level, a service encounter is one human being interacting with another. This interaction uses a good portion of a person's time in present-day society. However, when an individual chooses "service" as his or her profession, the amount of time required of the individual is multiplied many times over. *This added time should be spent in the development of a sound and proper "service attitude.*"

EXTRA EFFORTS REQUIRED BY A SECURITY OFFICER

Consumers and customers want satisfying experiences in their dealings with those providing security services. The desire for these satisfying experiences should be even greater on the part of the contractor responsible for providing these services. In order to accomplish this satisfaction the job of doing it often falls upon the shoulders of the security officer. Many times this individual is required to stifle personal feelings and control his or her emotions. Officers must find it within themselves to be able to direct behavior on the job in the direction their employer wishes. Contractors, through their officer employees, must satisfy the wishes of paying clients.

WHAT SATISFACTION BRINGS

If clients are satisfied with the services provided by the service supplier, and if they feel their business is valued by the contractor and treated accordingly, they will remain a client for a long time. In like manner, if clients do not feel satisfied, for whatever reason, they will switch contractors very quickly.

Whenever clients are doing business with anyone and are able to make a profit, they think long and hard before making a change. Clients will absorb a great deal more inconvenience from a profit-making venture than they will from a service that is costing them money and giving them no visible financial return on their investment. When security comes to being strictly overhead expense, the client's fuse is much shorter than normal. Security officers working for a contractor must know and understand this fact. Failure to do so can cost them their jobs and the job of the contractor.

The end result of keeping a client satisfied and happy is profit. Profit is what pays the salaries and the bills. If the client is not satisfied and happy today, that person won't be a client tomorrow. If many clients aren't happy tomorrow, the service company is out of business the same day.

WHO IS RESPONSIBLE FOR SATISFACTION?

Keeping a customer or client happy is not the sole job of the security company's management, nor is it that of the employees of the company who work at a client's place of business. Keeping a client happy and satisfied is the result of a combined effort on the part of both management and employees. If either one fails, usually the contract is lost. For some unknown reason, clients or customers seem to have little patience with service companies and don't hesitate to dismiss them. For each contract lost, profit is lost and employees usually suffer as a result. If all parties in the service company realize and understand their role and the important part they play, there is no need to incur customer or client dissatisfaction.

WHAT IS REQUIRED TO BE A SUCCESSFUL SECURITY CONTRACTOR?

To be a successful security service company, both sides—management and employees—must know and clearly understand the other side's role.

Management must

1. solicit the contract,
2. provide and train officers,

3. iron out and correct problems,
4. maintain an open line of communications with the client and with employees, and
5. be on the lookout for ways of improving service.

Security personnel must concentrate on all elements that contribute to customer satisfaction. These include:

1. proper attitude,
2. a neat and clean appearance,
3. knowing the job,
4. showing a desire to be cooperative,
5. willingness to give a little more,
6. knowing the client is the boss,
7. knowing things are done the boss's way,
8. being alert, and
9. willingness to help where needed.

The greatest and most important weapon against competition is having a happy client. A happy client is usually willing to sit down and discuss and negotiate in return for quality service and to stick with the company when times are difficult. Satisfied clients are often willing to allow for the fact new personnel require break-in time and to tolerate learning mistakes.

Providing superior customer satisfaction is the cornerstone of the approach to business. The difference between winners and losers is not what they do, but how they do it.

Differences Between Security Contractors

Few differences actually exist between security contractors. The most critical difference is found in the contractor who, throughout the entire company, emphasizes customer satisfaction. Success here will provide the greatest profit.

Successful companies do not leave customer satisfaction to their sales force while everyone else goes in different directions. Everyone works to satisfy and please the customer.

Customer-focused organizations make it everyone's business to know the customer in depth. Ranking high in customer satisfaction places the company in line for more business through "word of mouth" advertising by the customer. There is no need to sacrifice profit to make a customer happy. The only thing necessary is the willingness on the part of the service company and its employees.

The following are key characteristics of a successful security service company:

1. Set high standards.
2. Be obsessive about knowing what the customer wants.
3. Create and manage customer's expectations.
4. Design services to maximize customer satisfaction.
5. Make customer satisfaction everyone's business.
6. Do what you say you are going to do.

CHAPTER 4 PROBLEMS

Problem 1

Henry Swak is the CEO and chairman of the board of Swak Aircraft Corporation, one of the oldest aircraft manufacturers in the country. At 96, Mr. Swak is also one of the oldest board chairmen in the country.

Mr. Swak is known to have a very bad temper and takes out most of his anger on his employees. The lower the employee is on the ladder, the more that person seems to get.

Swak's facilities are large and they employ a large number of contract security officers. The sad part is that the security companies have been replaced four times in the last four years. It seems as though Mr. Swak finds something he doesn't like with each contractor.

Being a widower, Henry Swak works every night of the week into the late hours.

On Friday evening, guard Bert Simons is making his rounds through the executive tower when he sees a light coming from the open door of Mr. Swak's private office. This disturbs Bert considerably, but he doesn't want to give Mr. Swak the impression he is checking up on him.

After due consideration Bert feels he has a job to do and if Mr. Swak doesn't approve, Mr. Swak will just have to notify Bert's boss.

Bert forgets Mr. Swak's temper and peers inside the office. Mr. Swak is slumped over his desk. What should Bert do?

CHAPTER FIVE
Professional Relations

Visualize, if you will, the reception lobby of a major aircraft manufacturing company. The building in which the main reception lobby is located is immediately adjacent to one of the major runways of an international airport. We will call the aircraft company the Super Aircraft Corporation.

For over two years the sales staff of Super Aircraft has been negotiating a $5 billion contract for fighter aircraft for the air force of the Sultan of Squat. Squat is a major oil-producing nation. The Sultan has never been to the United States and insists on finalizing the contract himself at the offices of the aircraft company. All of the arrangements have been made by the company personnel for the arrival of the Sultan. The plant, offices, and the grounds have all been cleaned and spruced, and all of the company personnel have been trained regarding their conduct in the presence of the Sultan. They have been given to understand that the customs of the nation of Squat are considerably different from those of the United States. They have been cautioned they must be careful not to offend the Sultan by improper behavior. All was in readiness, but there was one error made in the preparations. No one remembered to tell the contract security officer who daily manned the reception desk in the main lobby that the Sultan would be arriving.

The officer who usually manned the post was Tom Smith, someone who was quite capable and who performed his job well. Tom also represented the aircraft company quite admirably. This day, however, Tom was nursing a bad hangover. The night before was his bowling team's banquet, and today he was in a very foul mood.

At the designated time, the Sultan's private jet landed and taxied onto the aircraft company's VIP ramp. The Sultan's guard alighted from the jet and formed a cordon from the plane to the lobby door. They all stood at attention with their ornamental curved swords gleaming in the sunlight.

Tom was writing in a log book when the Sultan strode into the lobby. Tom was so absorbed in what he was doing that he didn't realize anyone was standing in front of him. After a few embarrassing moments of being ignored, the Sultan loudly cleared his throat, hoping to get the officer's attention. Tom didn't even look up. He kept on writing and spoke out in a crusty voice, "Yo! What the hell do you want?"

This may be an improbable story, but it does underscore a very important point that all security personnel should be aware of. *The security officer may be the first person a visitor sees when calling on a client. The impression the security officer makes will most likely be the individual's impression of the entire company.*

PEOPLE WHO WORK WITH PEOPLE

The material contained in this chapter is written for those who work with other people. It's based on the assumption that, although the tasks and roles people play may vary, there is a foundation of attitude, knowledge, and skill basic to their work. Over a period of time, ideas and processes become such a part of an individual worker that they become that person's professional working self. The working self can hardly be distinguished from the personal self.

A FUNCTION FOR BOTH SIDES

There is a primary function or act required of a top-notch security officer. The function is "to serve the customer."

Because the "serving" is done by both the security officer and the service company management, this chapter is directed to the individual security officer, but it is equally applicable to management.

AN EFFECTIVE SERVICE ATTITUDE

The effective security officer must learn to avoid judging a client's attitudes and behavior according to the officer's personal value system. In the service business, the dominant value plays a very large part. The dominant value is the will and wishes of the customer or the client. It is the customer who pays the bills.

"The customer is always right" whether you agree with this statement or not. Regardless of your personal opinion of this view, it is the main theme upon which a successful business is built.

THE NEED FOR ADAPTATION

The service business requires a certain degree of adaptation on the part of service employees. Employees must adapt to the dominant value system. It makes no difference whether the employee has a different personal opinion. Until informed differently by management, the customer's opinion becomes the dominant value system.

Security officers must have sufficient self-awareness to be able to distinguish between value changes essential for good social functioning, and those changes dictated by their own value system. People's value systems are usually so deeply rooted they are often unaware of the reason for adopting and using them as a basis for judging effectiveness of behavior. This applies to security officers. Thus effective security officers must:

1. Be aware they are a walking system of values. These values are so much a part of them that they are hardly aware of their existence or the correctness of their values.
2. Use every means possible to learn about their own biases. Knowing and admitting biases is the first step in overcoming them. The system of overcoming biases is similar to the system used by Alcoholics Anonymous to overcome drinking.
3. Evaluate themselves and their values objectively and rationally.
4. Make every effort to change those values that need changing.

THE SECURITY OFFICER'S PERSONALITY

During their lifetime, security officers develop their own lifestyle, manner of thinking, feeling, and behaving. They develop personal values, a sense of identity, and a way of life. Each individual's differences are what set him or her apart from others and create a gap that must be bridged.

Life experiences which help establish ways of thinking and behaving are partially responsible for our personal development. As the worker reacts to the client, so too does the client react to the worker. Often the result is total incompatibility, which usually stems from the worker's difficulty in understanding those different from oneself.

A client who remembers a happy experience growing up with a blunt, outspoken, dominating, but still loving parent may easily relate to a worker with similar traits. Another client may shy away from such a person and never see the underlying good in that individual.

Security officers must be fully aware of their own personality assets and liabilities. They must learn to understand the effect they have on others. They must know themselves thoroughly and be prepared to change those personality liabilities that prevent them from having a meaningful relationship with others.

Successful people adapt to the changing demands of their own potential and their environment. If the security officer is to get ahead, he or she must be able to do the same. Once officers realize this fact they will do what is necessary to adapt as needed.

THE BASICS OF COMMUNICATION

When two or more people interact, communication takes place. Basically, we communicate what we are. For this reason, it is necessary to be aware of, and control or change, the destructive parts of our personality if we are to work with others.

To communicate there must be both a receiver and a sender.

The development and maintenance of communication with clients is a primary task of a security officer.

By definition, *communication is the transfer of meanings from one person to another.* A breakdown in meaningful communication is one of the great problems of our society. It is a cause for much of what is wrong in the world today.

There is much more to communication than simply speaking one to another.

The speaker (sender) evaluates the person to whom one wishes to communicate (receiver). Because the sender is unable to transfer his or her actual thoughts, attitudes, and feelings as they really are, the sender encodes them in a way he or she hopes best represents these thoughts, attitudes, and feelings. The listener (receiver) hears the message and translates or decodes it into a form the listener can understand. The receiver then encodes the response and sends it back to the sender. There are forms of interference that can distort or disturb communication of the message. These interferences can be personal feelings, social pressures, individual attitudes, or perceived realities.

In order for security officers to establish clear channels of communication between themselves and clients, they must remember the following:

1. Attitude of both parties is vitally important.
2. Understanding any similarities and differences between parties will determine the extent of understanding and acceptance of what is being said.
3. The capacity of persons involved to use both verbal and nonverbal communication, and to be able to interpret the symbols, is necessary.
4. The officer must be able to use a method communication that meets the understanding requirements of the client.
5. The officer must be able to determine when repetition and emphasis are necessary and be able to ensure feedback from the client to test client comprehension.
6. The officer must use gestures, symbols, and unusual words only as their meaning is commonly understood and accepted.
7. The tempo and level of communications should be geared to the level of the client, not the officer.
8. The officer must know how to adjust to compensate for interference as it arises, both in his (her) and the client's communications.

NONVERBAL COMMUNICATION

Nonverbal communication, or communication without the use of words, dates back to primitive times. It was one of the main ways for our ancestors to communicate with one another.

In normal communication between people, one-third is done by means of the use of words and two-thirds by use of symbols, gestures, voice inflection, or body language. Nonverbal communication is the main form used by those who do not have a good command of a language. Whenever strangers meet, nonverbal communication takes over in the beginning, and verbal communication increases as people get to know each other better and have evaluated one another.

Nonverbal communication can be continuous, with or without verbal accompaniment. Feelings and attitudes are principally conveyed in this manner in the early stages of a relationship. Nonverbal communication continues on till a lesser degree of familiarity takes hold.

There is always a danger of one person misunderstanding the nonverbal message of the other. Verbally one might say, "I'm glad you stopped by," but his or her body may be saying nonverbally by its actions, "I'm tired, and I wish you hadn't come."

Nonverbal communication is confusing if not understood, and it can cause a lot of problems. It can even be destructive to vulnerable, unknowledgeable individuals.

Security officers who have sensitivity are aware of their own feelings. They realize the impact their impression, through nonverbal channels, has on people with whom they work. Knowing this, they will make every effort to deal with people in constructive ways. If an officer has a problem and wishes to vent his or her wrath, that individual will choose someone who has a knowledge of him or her and who will encourage the ventilation of the wrath rather than encourage a negative attitude. This is where an intelligent, knowledgeable supervisor is needed.

Many times nonverbal communication may not be recognized, but it can have a great influence in making decisions and judgments.

All nonverbal communication is delivered through a person and an existing setting. Appearance, physique, posture, body odor, dress, tension, facial expression, behavior, silence or speech, tone of voice, gestures or movements, eye contact, touch, body sounds—all convey messages to the receiver. The physical setting does likewise. The appearance of the setting, the aesthetic quality, comfort, privacy or lack of privacy, and general climate also convey messages to the receiver. Once officers are alerted where to look, what to listen for, and what to sense in themselves and clients, their sensitivity to the client, and the client's ability to understand, will increase.

Examine voice tones. From the noncommittal tone designed to conceal, to the unrestrained expression of joy, pain, anger, fear, or grief, is a revealing part of conveying messages. Word meanings can vary greatly according to the tone of voice used when speaking them.

Messages are also conveyed vividly with the use of facial expressions. The facial expression is a powerful messenger. To a great extent, a person's expression becomes set by the life patterns of the individual. There is a uniqueness in the person's expression of apprehension, happiness, anger, passivity, friendliness, aggressiveness, and so on.

Silence is a very powerful form of nonverbal communication. It can express many things depending upon the context in which it is used. It can be interpreted as an expression of hostility and create anxiety. Because of this the security officer must be aware and knowledgeable of the client's understanding of silence if the officer is to employ silence as a means of communication.

Body movements and gestures have long been a principal method of expressing ideas and attitudes. One of the most frequently used is the use of one's eyes. Eye contact, depending on the duration and intensity, can be a significant factor in assessing the state of mind or feeling of the person with whom the officer is communicating.

Physical appearance sends a definite message about the state of mind and feeling of an individual. It also expresses a great deal about the person's thinking and personality. Conformity or nonconformity with generally accepted patterns of appearance carry a message of cleanliness, or the lack of it.

The overall physical appearance of people, the way they stand, the way they sit or lie down, sends a message to an observer.

WORD EXPRESSION

There is an old saying, "I say what I mean, and I mean what I say." This is not always true. In fact, often we do neither. To convey our intended meanings through words is a difficult process. There are differences in the meanings and uses of specific words. Differences exist in language and in the limitation of people's vocabularies. There is difficulty in expressing the subtle variations of meanings and feelings.

It may be difficult for some people to express an idea or feeling in words because they may fear the reaction of the listener. People cannot accept the idea of struggling to find the proper word or expression.

When such a condition exists the speaker may disguise what he or she is saying in such a way that the true meaning, although actually present, is hidden among the verbiage. If what needs to be said or expressed is too painful or threatening, the speaker's overreaction may be great enough to say just the opposite of the intended meaning.

The ability to speak honestly and clearly is dependent upon selection of words, the perceived and understood meaning of these words, the context in which they are used, and any specific meaning given them by the culture of the sender and the receiver.

INTERPRETING THE MESSAGE

Regardless of the medium of communication used, the receiver interprets the meaning of the message. Because people are different in many ways, these meanings do not always result in the understanding of the intended thought or feeling.

Three things are necessary to remember when discussing the interpretation of messages: perception, culture, and context.

Perception

Perception is the way in which communication is understood intellectually, how it is seen and understood through the senses, and what it actually means to the receiver. A communication can be understood to mean something entirely different from the meaning intended by the sender. What might be construed to be a glare of rage could be merely the grimace of an anxious person unsure of himself and what he is doing.

When the receiver becomes aware of the communications sent on the same level from which they are sent, the channels are open and operate effectively. An understanding is achieved. When the level is different, there is a breakdown in communication.

Perception is strongly influenced by the cultural background of the one receiving the message. The receiver's life experiences could alter the perception. For these reasons a person who wishes to communicate clearly should keep both facts in mind.

Culture

There isn't a security person anywhere who is able to learn all of the cultural variations with which one may encounter in a working lifetime. However, security officers can help themselves considerably if they are aware of and open to cultural differences. Officers must not be defensive about their own attitudes and behaviors, and they should have the ability to say to others and to themselves, "I don't understand" or "Do I read you correctly?"

Context

The total context in which messages are conveyed is still important in the determination of their meaning. The security officer must not only look at the expression, but also at the message as a whole, both with respect to the time and place of its expression. The message that may seem quite inappropriate in a specific form of reference may be perfectly logical and understandable in a different one. Most school teachers are familiar with the youngster who arrives at school in a destructive or overactive mood, which is in direct contradiction to the normal, acceptable manner of behavior. The teacher also knows the child is reacting to something that happened outside of the classroom and that the youngster is making his or her feelings about the incident known.

INDIVIDUAL RIGHTS

Every individual has a right to exist, to have importance and value. The basis of any relationship is the acceptance of these rights of each person by all parties in the relationship. Indifference to any or all of these rights can completely prevent or destroy the development of a meaningful relationship. The results can be more destructive than merely disliking someone. Like and dislike are dynamic feelings that usually precede an action of some sort. Indifference, however, is a condition from which nothing grows. From the acceptance of others should come freedom to be oneself, to express one's fear, anger, joy, rage, to grow and develop and change without being concerned that such actions will put a relationship in jeopardy.

Acceptance also recognizes the uniqueness of the individual. A person possesses the right and the need to participate in making decisions affecting one's own welfare. When people use their right of self-determination to affect adversely the rights of another person, there becomes a need for some control and limitation. Individual rights do not give one person the right to infringe on another person's rights. Individual rights do not negate the rights of both parties in a friendship or relationship or the existence of the need for some control and limitations.

People often confuse acceptance with "liking" an individual and approving of his or her behavior and of the individual as a person. Likes and dislikes are products of acceptance. An important part of the development of a good relationship with a client is the genuineness of the service person's acceptance and concern for the client.

We not only accept people for what they are but for what they are capable of being. The officer's expectation of the client or company's potential will influence his or her attitudes and behavior. It has a definite bearing on the outcome of their joint interactions.

CHAPTER 5 PROBLEMS

PROBLEM 1

"A security officer who displays a/an ______ that denotes an overbearing demeanor will seriously damage a favorable image for security personnel." The blank space in the above sentences refers to:

A. Appearance
B. Attitude
C. Visible knowledge
D. Bearing

PROBLEM 2

When a citizen asks you to give information you have been entrusted not to make public, you should,

A. Consult your supervisor.
B. Inform the citizen to keep the information confidential.
C. Tell the citizen you are not permitted to reveal such information.
D. Tell the person such information is for security personnel only.
E. Advise the person that, since you are only a patrol officer, you do not feel you have the authority to give out the information.

PROBLEM 3

Which factor has an important effect on the prestige and public relations of a security agency?

A. Contacts with visitors.
B. Telephone contacts.
C. Correspondence.
D. Administrative ability.
E. All of the above.

PROBLEM 4

Security officers can deal effectively with the public only if they consider each person as

A. an individual.
B. economically essential to them.
C. equally important.
D. active in a distinct environment.

PROBLEM 5

Security Officer Jane Doe is working as a "fill-in" during the lunch hour for the regular receptionist at the Jones Manufacturing Company.

A very disgruntled person comes through the front door and walks up to the desk. Without so much as a "good morning," the person addresses Officer Doe. "Young lady, what does your company think people are—camels? It's ninety-eight degrees outside and I had to walk over at least 200 yards of hot asphalt parking lot to get to the front door. Why don't they make provisions for visiting customers to park closer to the office?"

How do you think Officer Doe should handle this situation?

Chapter Six
Arrest

In the course of their normal duties, private security personnel may on occasion be confronted with an arrest situation. It is at this moment when it is absolutely necessary for the security officer to know and understand exactly what constitutes an arrest, who may make an arrest, and for what an arrest may be made.

WHO MAY MAKE A LAWFUL ARREST?

The following individuals may make a lawful arrest:

1. Federal, state, and local law enforcement officers
2. Peace officers
3. Security officers
4. Private citizens

Although this may sound like a statement giving each of the above parties the same authority, it is not. There are clear-cut differences in when and how each of the above parties may accomplish a legal and lawful arrest. We will do what it takes to make the distinctions clear, but readers must always remember to check out the arrest authority of each party to make an arrest in their individual states.

VARIATIONS IN THE ARREST AUTHORITY OF SECURITY PERSONNEL

The arrest authority of security personnel varies through the country. Each state in the union has the right to regulate the dispensing of arrest authority within its own boundaries. Once again, because of this fact, we remind readers to check the laws of their own state.

We will make every effort to write about this subject generically and to let each student find his or her own place in the scheme of things.

As of 1988, there were 38 of the 50 states that required statewide licensing of security personnel. By that is meant the state, not a county or city, regulates the authority granted private security and proprietary security officers. This includes not only the authority granted but also statewide security rules and regulations as well as uniform regulations. The security officer, under statewide licensing, may work anywhere in the state in which he or she is licensed without paying an additional county, city, or municipal licensing fee.

In those states in which statewide licensing is not in effect, the authority for the security officer's powers, including arrest, is granted by the state's county, city, and municipal judicial representatives.

Each individual officer must be certain of his or her own status in the state in which one is licensed. In states that do not have statewide licensing, the legal authority may change from one city to another within the state. Many counties within that state may not have any licensing procedures; in this case, doing security work under those conditions is to be doing so as an ordinary citizen.

Summarizing these general guidelines, the reader should consider the following:

1. There are times, in certain states, when security officers have the same authority as a police officer within the confines of a client's property, and subject to certain conditions.
2. There are times, in certain states, when security officers have no more authority than that of a private citizen.

There being no single rule to follow for all security officers, we will begin with a discussion of arrest (regardless of who makes it), the arrest procedure, the elements of arrest, and the consequences of arrest. Everyone who is qualified to make an arrest should know as much as possible about each of these phases regardless from where the authority derives.

ARREST: WHAT IS IT?

An arrest is a procedure whereby a citizen, with legal authority, deprives another citizen of his or her freedom and liberty as guaranteed under the Constitution of the United States.

The term *arrest* "signifies the apprehension or detention of the person of another in order that he may be forthcoming to answer for an alleged or supposed crime" (*State* v *MacKenzie,* 210a, 2d, 32–33, Me., 1965).

This action, an arrest, *is an act of the most serious nature and has far-reaching effects that remain with those arrested for the remainder of their lives.* For this reason, each state, in its statutes, clearly defines and strictly enforces how and when the authority for arrest is invoked.

WHAT ACTIONS ACTUALLY CONSTITUTE AN ARREST?

This question is one of extreme importance. Every security person must know and comprehend the entire answer. In order for officers to assume their proper place in the criminal justice system, it is of the utmost importance to officers, their employers, and their employer's clients to *understand fully all of the immunities, responsibilities, and liabilities of making an arrest.*

A person is arrested when, by the actions or words of another, the person arrested is no longer able to go or act within his or her own discretion, or when the person is no longer free to go at his or her own pleasure.

Moreover, *there need not be physical force to constitute an arrest. It is also unimportant whether or not the person, or persons, arrested are ever prosecuted. The state of mind of those being arrested or believing they are being arrested is an important factor.*

> The mere touching of the person is also considered to be an actual seizure and may constitute a formal arrest, if the other elements of formal arrest are present. (*Childress* v *State,* 175, A2d 18 Md., 1961)

> The peaceable submission eliminates the need for physical action and satisfies the requirement of seizure or detention. (*State* v *Donahue,* 420, A2d 936. Me., 1980)

When an officer, in his or her official capacity, uses force on a citizen, the officer has arrested that citizen. Force, seizure and restraint, in and of themselves, are actual or constructive control of a subject by the police, peace officers, security officers, or, technically, private citizens. The use of force on a citizen is the strongest irrefutable evidence of an arrest.

The law of arrest, all across the country, holds that almost any restraint on the part of a peace officer, police officer, or security officer, regardless of how slight, coupled with the intent to arrest, constitutes an arrest.

ARREST AND ITS CONSEQUENCES

An arrest by any lawful authority takes away a person's freedom and places that individual in the custody of the person making the arrest or the custody of the law. It can be the beginning of imprisonment for the person arrested.

The issue of whether or not an arrest has been accomplished is of great importance. It holds a great deal of significance in both criminal and civil law.

Anyone making an arrest may be liable in civil suit for a mistake if the person arrested is not guilty of criminal conduct.

In civil actions for false arrest any display of authority will increase the amount of damages assessed because of the additional embarrassment caused the subject by the false arrest.

Citizens are not required, under ordinary circumstances, to surrender their liberty upon an arbitrary demand of a law enforcement office, peace officer, security officer, or an ordinary citizen.

The U.S. Supreme Court has ruled that

> Law enforcement officers do not violate the Fourth Amendment by merely approaching an individual on the street, or in another public place, by asking him if he is willing to answer some questions, but putting questions to him if the person is willing to listen, or by offering in evidence in a criminal prosecution, his voluntary answers to such questions.... Nor would the fact that the officer identifies himself as a police officer, without more, convert the encounter into a seizure requiring some level of objective justifications.... The person approached, however, need not answer any questions put to him; indeed, he may decline to listen to the questions at all and may go on his way.... He may not be detained even momentarily without reasonable, objective grounds for doing so; and his refusal to listen or answer does not, without more, furnish those grounds.... If there is not detention, no seizure, within the meaning of the Fourth Amendment, then no constitutional rights have been infringed. (*Florida* v *Royer*, 460 U.S. at 497098, 1036 S.Ct. at 1324, 75L.Ed.2d at 236 [1983])

According to approved practice, the person about to be arrested should first be informed of the purpose of the arrest. It has been held by some courts (*State* v *Phinney*, 42 Me.384) "if this is not done, resistance on the part of the person being arrested is justifiable. This is not true in all cases, and should not be accepted as a general rule. Circumstances surrounding the arrest has a bearing on the issue. Circumstances may require extreme caution on the part of the arresting officer when the person to be arrested is known to be a dangerous felon who is likely to resist and endanger life. In such a case, public policy would seem to dictate concealment of the purpose of the arrest until appropriate decisive action on the part of the arresting officer or officers renders the subject into complete custody."

It should be constantly in the mind of any person making an arrest that the person arrested and booked, even though never charged or brought to trial, is forever marked. The arrest record stays with that individual the rest of his or her life.

ELEMENTS OF CRIMINAL ARREST

Four essential elements constitute a criminal arrest:

1. an *intent* by an officer to make an arrest;
2. *authority* to make an arrest;
3. a *seizure or restraint,* actual or imagined in the mind of the person being arrested;
4. an *understanding* by the person being seized that he or she is being arrested.

In the "criminal arrest" as compared to the "civil arrest" situation, *all four critical factors must be present in some degree.*

The *intent* of the officer making the arrest to take a person into custody is the element of arrest that separates the formal arrest from lesser forms of restraint or detentions. Some of these forms are:

1. momentarily stopping a person to seek information or ask assistance;
2. requesting a witness to appear at the police station for the purpose of an interview;
3. stopping a vehicle to inspect it for proper licenses or load limits;

4. serving a subpoena;
5. restraining someone who is violent or behaving in a manner that causes danger to the person or others nearby.

The first vital issue in an arrest is the *time the arrest occurred.* At this point the rights of the person suspected or being detained change. *It is at this point where custodial responsibilities take over for the arresting party.* At this time the person arrested is no longer free to flee from custody or to resist arrest without incurring further penalties.

In a suppression hearing of the court on a warrantless arrest-search situation, *the court has the sole responsibility to determine factually, from the circumstances of the case, if there was actually an arrest, and assuming there was, the exact moment the arrest took place.*

In its effort to make a proper and just decision, the court looks to see *what the officer's purpose was, what he or she knew and observed, and exactly what the officer did.*

Of the three determinants to an arrest mentioned in the preceding paragraph, the most important, from a practical point of view is, *what the officer did.* What the officer did is the focal point of his or her purpose and what the officer knew or observed. *It takes the act to complete the arrest.* This action is the only thing that can complete the arrest, although alone, without purpose or knowledge, the arrest would be illegal and of no value.

POWERS ATTACHED TO ARREST

With the "act of arrest" comes these arrest powers:

1. use of force
2. right to search
3. right to seize or restrain.

Any time a law enforcement officer, security officer, peace officer, or private citizen uses one of these arrest powers he or she rings the *constitutional warning bell.* The officer calls attention to oneself and the incident. Very close scrutiny becomes the watchword. The user of any such power had better be able to stand close scrutiny of one's actions and behavior. There is a tremendous amount at stake in the outcome. Seldom are compliments issued, but there are often complaints. Often, charges of one kind or another accompany the complaints.

WARRANT ARRESTS

Normally, except under extreme conditions, security personnel will not be involved in warrant arrests. This emerges from the fact that the warrant arrest is part of the judicial process, and warrants issue only to law enforcement or judicial officers.

Deputization of security personnel as law enforcement officers is done only under the most extreme conditions, if ever, and these conditions normally do not include the serving of a warrant or arresting with a warrant.

WARRANTLESS ARRESTS

Warrantless arrests can be made by any peace officer, law enforcement officer, security officer, or private citizen. The decision as to who may make what types of arrest is left to the discretion of the individual states.

In most states the arresting party can arrest for a felony without a warrant if he or she has the required probable cause. Most arrests of this sort are sight arrests and are made without the obtaining of warrants.

For years states contended an arrest for a misdemeanor could only be made with a warrant, unless the offense was committed in the presence of the police officer. Most states have reviewed this position, and at this time most police officers can arrest for a misdemeanor case on probable cause without a warrant. In some states, full arrest authority is also given to security officers. While on duty on the client's premises, he (she) would have same arrest powers of a police officer. Since this may not currently be the status in the reader's state, *it is highly recommended you check your state statutes and laws concerning arrests for misdemeanors.*

Police or law enforcement officers, be they federal, state or local, may make warrantless arrests under the requirement known as "probable cause." Probable cause for an arrest is defined "as a combination of factual facts, or apparent facts, viewed through the eyes of an experienced police officer, which would lead a man of reasonable caution to believe a crime is being, or has been committed" (*Brinegar* v *United States,* 338, U.S.160,69S.Ct., 1302,93L.Ed.1879 [1949]).

The authority of security officers to make warrantless arrests varies by state. Because of these variations, security officers should make sure they know their authority, from where it comes, and when they can use it.

In the beginning of the chapter we mentioned we would discuss each of the four categories of individuals empowered to make arrests. This discussion follows.

1. Law enforcement officers are permitted to make both warrant and warrantless arrests at any time, provided they have "probable cause."
2. Peace officers, (usually sheriffs and their deputies, constables, and marshals) usually have most of the authority of law enforcement officers, depending entirely upon the laws and statutes of their state.
3. Security officers' authority to make warrantless arrests is contingent upon the laws and statutes of their individual states. In many states where licensing is the rule, security officers have the authority to make arrests for crimes committed in their presence as long as they are within the confines of a client's property.

   In states where there is no state licensing, the authority to make warrantless arrests by security officers is left to the individual county judicial authority, city judicial authority, or municipal judicial authority. No matter who issues the authority, it is normally only authorized for use within the confines of a client's property while the officer is on duty.

   If the officer is performing in a particular county of a state that has not prescribed authorization for security personnel, the officer is then usually only to use the authority given a private citizen.
4. Private citizens have the authority to arrest anyone committing a crime in their presence. However, making a citizen's arrest leaves the individual making the arrest vulnerable to numerous civil lawsuits.

EXAMPLES

The following examples concerning the authority of security officers are taken verbatim from the manuals of a large midwestern city's security officer's manual, and from a specific county of a midwestern state. These examples are from states that do not have statewide licensing and where such licensing is left to the principal judicial authority in the city or the county. Similar situations may exist in other parts of the United States. *Once again, it is important for each security officer to seek out the governing authority in the area wherein he or she works to determine exactly what the officer's authority is.*

The City Security Authority Example

1. The oath, prior to issuance of his/her license, an applicant must swear to uphold the following;

   I do solemnly swear that I am a citizen of the United States, or a legal resident alien, that I will faithfully support the Constitution of the United States, the Constitution and the Laws of the State of ____________ ______________________and the Charter and City Ordinances of the City of ______________________ ______________________; that I have never been discharged from the police force of the City of ________________________________; that I have never been convicted of a felony; that I have no physical or mental disability or habit that disqualifies me from performing the duties of a Private Security Officer; that I will wear such dress, badge/ID card or emblem as the Board of Police Commissioners may from time to time designate; that I will, to the best of my skill and ability, diligently and faithfully, without partiality or prejudice, discharge my duties according to the Constitution and Laws of the State of ________ __________________________ and Charter and Ordinances of the City of ________________________ ______; that I will strictly obey all lawful orders and regulations of the Board of Police Commissioners of the City of ________________________________, the Chief of Police, or any officer placed by them over me; that I will not cease to perform my duties until my resignation is accepted by the Board of Police Commissioners; that I will not become a member of, or affiliate myself with, any organization of any kind or character whatsoever, membership which will or may impose upon me obligations inconsistent with the full performance of my duties as a Private Security Officer, or inconsistent with the oath herein taken to carry out the orders of the Board of Police Commissioners and to comply with its lawful orders, rules and regulations, or which will or may, in any degree interfere with the performance of my duties as a licensed security officer.

AUTHORITY

Purpose

This rule establishes the arrest powers of a licensed security officer. These powers are in effect only while the licensed security officer is in the area designated by his/her employer and during the time he/she is assigned to work.

Authority

Private security officers have the authority to make an arrest and to search for and seize evidence in connection with the arrest, at the location and during the time of their assignments, under the same conditions as members of the police force of the City of ______________________________.

A. In all instances of felonies, misdemeanors and city ordinance violations committed in the presence of the officer.

B. During an attempt to commit a felony or a misdemeanor.

C. For an offense not committed in the presence or view of the security officer, when he/she has probable cause to believe that the offense was committed by the person she/he is arresting, and

D. Off his/her licensed premises when in hot pursuit for an on-view felony is involved. (An on-view felony offense is an offense the security officer sees committed.)

E. The authority granted private security officers is limited to designated areas only, and does not include such services as bodyguard, escort, process server, or investigative service for lawyers in criminal or civil activity.

The County Security Manual

A. Authority

1. A watchman or courier shall not have the authority to make arrests, physically detain persons, or conduct search and seizure.
2. Licensed security officers have the authority to make an arrest and to search and seize evidence in connection with the arrest, at the location while on duty for:
   a. Any crime committed in their presence,
   b. For a felony not committed in their presence or view when there exists reasonable grounds based upon probable cause, to believe that the offense was committed by the person he is arresting.
3. An arrest requires that the defendant be restrained or that the defendant submit to the custody of the officer for the purpose of bringing him before the court.
4. An arrest will be deemed to have been made if a person accompanies an officer because he reasonably believes that the officer intends to use force, if necessary, to control his actions. If the officer does not intend that an arrest be effected, he should clearly make this known to the person he is asking to accompany him.
5. In making an arrest, no more force is to be used than is necessary for overcoming any resistance that may be offered and for ensuring the safe delivery of the prisoner into custody. Abuse of prisoners, by word or by act, is prohibited.
6. When an arrest has been made by a licensed security officer, it shall be his duty to notify the police department in the jurisdiction where the arrest has been made.

B. Duties

General

1. It shall be the duty of every licensee to assist all police officers in the ________ County area in preserving the peace or in taking such other action as may be necessary to effect an arrest during the times and in the area where he is employed, and when requested to do so by a police officer.
2. Should a serious accident or crime, including all felonies, occur on the premises of the licensee, it shall be the responsibility of the licensee to notify the appropriate police department immediately. Failure to do so is a violation of the provisions of this manual.
3. The licensee will be responsible to protect the crime scene until relieved of that responsibility by the police department in charge.

As is evident by the previous examples, the authority granted some private security personnel is quite extensive. Those individuals having been granted such authority must know and understand the laws of their jurisdiction. To carry out the duties mentioned in the examples, the security officer should be well versed in the laws of arrest, search and seizure, and the use of force.

Note: The material in this chapter should be studied and understood by all persons having the authority to make arrests. Even though the arrest powers for security personnel throughout the country may differ, it is desirable to know as much as possible. Again, check the authority in your own state, and abide accordingly.

PROBABLE CAUSE

Before any of the arrest powers mentioned earlier in the chapter are invoked, *justification* or *probable cause* for using such action must be present. It is too late after one of the powers has been invoked to seek probable cause or justification.

> Officer Anthony Martin, while patrolling his beat in uniform, observed Sibron from 4 p.m. till midnight. During the period he saw Sibron in conversation with six or eight persons who he knew from experience to be addicts. He did not overhear conversations or see anything pass between Sibron and the others. He later observed Sibron in a cafe talking to a few more people. Martin summoned Sibron outside the cafe and said to him, "You know what I want" and when Sibron started to put his hand in his pocket, Officer Martin reached into the pocket and withdrew some plastic bags containing cocaine... (*Sibron* v *New York,* 392 U.S. 41,88 S.Ct. 1889, 20 L.Ed. 917 [1968])

The U.S. Supreme Court ruled "an incident search may not precede an arrest and serve as part of its justification."

The search was ruled illegal because there had been an arrest, and by the actions of Sibron prior to the search, there was insufficient probable cause for an arrest if there had been no search.

Probable cause to arrest exists where the facts and the surrounding circumstances, of which the arresting officer has reasonably trustworthy information, would justify a person of reasonable caution in believing an offense had been committed, and that the person arrested committed it.

NO PROBABLE CAUSE AFTER THE FACT

An arrest cannot be justified by the results of a search after an arrest (see the previous case, *Sibron* v *New York*).

A police or security officer may consider the past record of a suspect, and hearsay concerning the commission of a crime, even though both may not be admissible at a trial. Standing alone, however, such evidence would not be considered enough. You cannot arrest someone simply because that individual was once arrested or someone said that the person committed a crime. Courts have recognized that a trained police officer may often have probable cause to arrest for a crime based on facts and circumstances that would not produce probable cause in the mind of a security officer or an untrained layperson.

WHEN PROBABLE CAUSE CEASES

At the moment of arrest, probable cause ceases to build. Any facts that develop after the arrest will not be considered by the court as part of the probable cause needed for the arrest.

The time an arrest occurs is the first critical step for the prosecution in establishing probable cause for the arrest. Of the four elements necessary to determine whether there was an arrest and when it occurred, the element of intent and the element of seizure or restraint are the most important. These are the elements that give the court more direction than anything else.

The purpose of the arresting officer, in an arrest situation, is determined by very critical examination of the officer's actions, what he or she said and did, from the moment the officer confronted the accused until the moment it become perfectly clear, beyond any doubt, that the accused was in custody.

Time of arrest is dependent largely upon the purpose of the officer as shown by his or her words and actions. These acts are the key by which the court determines when an arrest took place.

THE VIEW OF CIVIL COURTS

Civil courts are extremely liberal and broad in their interpretation as to whether a person's conduct, words, or tone of voice placed that individual in the state of arrest. These courts lean more to an arrest taking place with much less effort on the part of the arresting party. Using their broad and liberal interpretation, these courts tend to view the total event *subjectively as to the state of mind of the person arrested* and *not objectively as to what was intended* by the acts and words of the arresting person. *If the court finds it was reasonable for subjects to infer from the arresting person's actions and/or words that they are arrested, then the arresting party is deemed to have perfected and executed an arrest.*

The theory behind the differences of arrest in civil and criminal matters seems to come from the fact that, in criminal matters, time out of a person's life span can be at stake, whereas in a civil matter a loss of an amount of money by the loser in the law suit is the result.

THE LIABILITIES CONNECTED WITH ARREST

The fact that the subject of the arrest is detained only a short time makes absolutely no difference to the court. The moment the subject is deprived of his or her freedom and the arrest takes place, the person depriving the subject of said freedom is fully responsible for all consequences that may follow as a result of the arrest.

Any time a person is arrested by another, the person doing the arresting assumes a number of obligations to the person arrested. These obligations are not the same under all circumstances; the difference depends on whether or not the one doing the arresting has the recognized status of an ordinary citizen.

The court places an obligation on anyone making an arrest for any reason to provide for the personal safety of the individual arrested. This responsibility and liability begins for the person making the arrest at the moment the arrest is made.

SURRENDER AND NOTICE

No one has a legal obligation to surrender voluntarily to arrest. It is not a violation of the law to avoid arrest. Once an arrest has been made, however, any act of avoidance on the part of the person arrested could be deemed "resisting arrest," which is a violation.

Most states require, by statute, the giving of notice in criminal matters. The notice takes the form of a warrant and requires the arresting party to make known to the subject to be arrested, not only of the intent to arrest, but also the reason for the arrest. There are exceptions, however, such as when the giving of notice may compromise the success of the arrest. There are times, such as a "crack house raid," when due to the need of a speedy arrest to avoid escape, injury, or destruction of evidence, the arrest is made without a warrant and then application for the warrant is made within a certain time span set by statute.

RETAIL SECURITY ARRESTS

Questions of concern often arise when an employee of a company is interviewed by the security personnel of a company at the place of employment. While the party is being interviewed, he or she is on company time. Some believe that as long as an employee is paid for time spent being interviewed, an arrest or detention cannot exist. *This opinion is wrong.*

A key question must be asked in order to address this matter. *Is the employee totally free to leave the interview any time during its duration without penalty of any kind?* If the answer to this question is "Yes, the employee can leave at any time," there is no arrest. However, if by the words or actions of the interviewer or others present the subject really believes that he or she is not free to go, the mere payment of wages to the subject does not exclude those doing the interviewing, or others, from having made an arrest or detention.

Obvious acts, such as blocking the only exit to the interview area or placing the person being interviewed in a chair and standing directly in front of him, put the subject being interviewed in a state of mind that says one is not free to go at one's leisure.

GENERAL AREAS OF LAWFUL ARREST

There are two general areas of lawful arrest. These are *arrest with a warrant, and arrest without a warrant.*

ARREST WITH A WARRANT

Since this book deals with the training of security personnel, and since they scarcely, if ever, become involved in a warranted arrest, we will not go into a lengthy discussion of the subject. Suffice it to say that all of the preceding material covers arrest in general and is necessary for a security officer to know. The warranted arrest, however, is primarily handled by police or other authorized law enforcement officers and is not required in this instruction.

THE WARRANTLESS ARREST

The authority to make a warrantless arrest is generally defined by state statutes. The majority of states, if not all, under varying situations or conditions grant law enforcement officers, peace officers, security officers, and private citizens the right to make a warrantless arrest of anyone committing a felony in their presence. Authority under other conditions depends on the statutes and jurisdictional authorities in the individual states.

NEED FOR PROBABLE CAUSE

In order to make a legal arrest, there is a requirement that is absolutely necessary before the arrest can be effectuated. This applies to both law officers and private citizens. In legal terminology it is known as "having probable cause." To define this to the satisfaction of everyone is impossible. There are times when the decision as to whether or not there was sufficient probable cause to make an arrest is not finally determined until there is a ruling by the U.S. Supreme Court.

The Fourth Amendment of the United States Constitution provides that

> The right of the people to be secure in their persons, houses, papers, and effects, against unreasonable searches and seizures, shall not be violated, and no warrants shall issue, but upon probable cause, supported by oath or affirmation, and particularly describing the place to be searched, and the persons or things to be seized.

The standard judicial definition of probable cause, at least as it relates to the arrest of a subject by an officer of any kind, is:

> Probable cause exists if the facts and circumstances known to the officer, at the time of the arrest, warrant a prudent man in believing an offense has been committed. (*Brinegar* v *United States,* 338,U.S.160,69,S.Ct.1302, 93 L.Ed.1879, 1949)

There are instances in which law enforcement officers (and legally authorized security officers acting within the confines of a client's premises) are permitted to make arrests without first obtaining a warrant. In such instances, it is the Fourth Amendment protection against

> unreasonable searches and seizures which is directly applicable. But, because a principal incentive for the procurement of warrants would be destroyed if police needed less evidence when acting without a warrant, the requirements in such instances surely cannot be less stringent than when a warrant is obtained. (*Wong Sun* v *United States,* 371,U.S.471,83 S.Ct.407,9L Ed.2d,441, 1963)

THE SECURITY OFFICER AND THE PRIVATE CITIZEN

Under common law, peace officers and private citizens could arrest without a warrant when a felony had been committed and there was reasonable grounds to believe the subject had committed the act. In most states these laws have been refined, and redefined. Though a citizen's arrest is still possible, people should know and understand the fact that in most situations where such an arrest takes place, peace officers enjoy a much greater immunity from arrest suits in that they will not be liable if it can be shown there was reasonable cause to believe a felony had been committed. (This is true also for legally authorized security officers while acting within the confines of a client's property.)

In a great many states the private citizen is not afforded this protection, and if in fact a felony was not committed, the arrest by the private citizen becomes invalid and liable to a lawsuit.

Security officers must be guided in matters such as these by the authority vested in them by the agency extending their authority in the jurisdiction in which they work. (Make certain you check your authority status with the authorizing agency within the state or jurisdiction in which you are employed.)

Security officers are usually granted full powers of arrest in any incident that takes place within the premises of the client for whom they are assigned to work, and have that power until the police of the jurisdiction arrive on the scene. At that time the security officer defers to the police and places himself (herself) at the disposal of the police.

A number of states have changed their statutes so peace officers can arrest without a warrant in misdemeanor cases that do not take place in their presence. In most cases security officers are granted the same type of authority in misdemeanor cases as they are in felony cases, *always in accordance with, and at the pleasure of, the authorizing agency or jurisdiction from which they are licensed.*

Remember, *all security officers should be quite certain of their status, their powers, and their authority before they begin work.* These may vary from state to state; thus it is your duty to investigate these matters thoroughly. You should fully understand the exact degree of your authority. If security officers are found to have a status comparable to a peace officer under certain conditions, they will also inherit all of the obligations and responsibilities of the police officer.

CONSTITUTIONAL RIGHTS AND OBLIGATIONS OF SECURITY PERSONNEL

Most constitutional rights we hear of relate to the rights of citizens with regard to their state and federal government. As a general rule, one citizen cannot violate the constitutional rights of another citizen. These rights may be violated only by state or federal law. It is possible, however, to violate a wealth of local and federal statutes relating to civil rights, unfair or discriminatory employment practice, invasion of privacy, labor laws, disclosure acts, etc.

MIRANDA RULING

Private security personnel effecting citizen's arrests today are not bound by the *Miranda* decision and therefore are not obligated to extend any warnings to persons they arrest.

As a result of two Supreme Court decisions, *Escobedo* v *Illinois,* and *Miranda* v *Arizona,* a peace officer must immediately inform an arrested person of his or her constitutional rights when being arrested for the commission of a felony, or when the person is placed under arrest following an interview in which the individual is suspected of committing a felony.

See *Escobedo* v *Illinois* (378 U.S.478,84S.Ct.1758,12L.Ed.2d 977 [1964]) and *Miranda* v *Arizona* (384 U.S.436,86S.Ct.1602,16L Ed.2d694 [1966] reh.den. 385 U.S.890,87.S.Ct.17L).

CHAPTER 6 PROBLEMS

PROBLEM 1

Shopping is heavy during the current sale at Hooperman's Dry Goods Company. It's almost like Christmas, and Security Officer Jim Adams has had his hands full. He knows he can't catch everybody, so he hopes the other two plainclothes guards in the store will do their share.

Jim has his eyes trained on a young man about 20 years of age who is acting very nervous, looking from side to side and over his shoulder. It's as though he's ready to do something he may regret later. Jim stays with the young man, keeping him in view. It doesn't take too long before the suspect is at the watch counter looking at the various timepieces on display boards resting on top the counter. They aren't the most expensive watches, but they aren't the cheapest either.

The young man keeps eyeing the salesperson who is occupied with another customer. Glancing back over his shoulder again, he places the watch he has been looking at in his left hand and moves to another rack to inspect another group of watches. Jim Adams follows behind him, acting as if he too is looking at jewelry.

While Jim is examining some rings, he looks up and notices the young suspect is walking away from the counter and is already turning the corner around some of the displays on the floor. Jim realizes he'll lose the suspect if he doesn't cut him off, so he takes a shortcut and finds himself in a position even with the suspect. Jim estimates the man could not have been out of his sight more than five seconds.

He watches the suspect go past the cashier at the checkout counter without stopping. Jim follows the man outside and stops him in the parking lot. Jim tells the man he'd like him to return to the store with him; the man says no. Jim then places the man under arrest.

Did Officer Adams have probable cause to make the arrest? Should he, or should he not, make the arrest?

PROBLEM 2

Marsha Williams had only had her new assignment as a security officer at Stacey's Exclusive Women's Shop a few weeks. She felt very professional in her cleaned and starched uniform, along with her leather belt, holster, handgun, and handcuffs. All was going well on her new job and there was no doubt of her visibility. The company had requested a guard who would stand out and act as a deterrent to theft.

Marsha was passing the open door of the store manager's office and heard the manager in a loud discussion with someone who appeared to be a patron of the store. The manger was saying, "Mrs. Smith, would you mind remaining here until we can clear up this matter?"

"No, I won't stay here!" Mrs. Smith said and she headed for the door.

Hearing this, Officer Williams stepped in front of the doorway and stated to Mrs. Smith, "You heard the manager. You'd better listen to him!"

Did this constitute an arrest? If so, why? If not, why?

PROBLEM 3

Guard Martin Adams is working on an assignment at the experimental unit of the World Atlas Company, a manufacturer of highly classified computer chips. There has been a good deal of industrial theft and espionage lately, and the management of the company is nervous. They only recently gave specific orders to all security personnel to be on the alert.

There is a free-standing structure located about 20 yards from the laboratory building. Over the entrance is a clearly written sign: NO ADMITTANCE—AUTHORIZED PERSONNEL ONLY.

As Adams is on his rounds, he sees an unidentified person starting to enter the door of this separate building. Adams runs up behind the individual, puts his hand on the person's shoulder, and yanks him backward away from the door. "Can't you read? That sign says authorized personnel only. Where do you think you're going?" Adams demands of the person.

Was or was not this an arrest? If so, why? If not, why not?

PROBLEM 4

Security Officer James Peak is working the aisles of the Meadow Brook Mall the Saturday before Christmas. It seems like the cold weather, plus the time of the year, has brought all of the shoppers indoors.

Crowds are finding it difficult to maneuver through the pedestrian traffic, but for the most part patience seems to be in command. Everyone is taking the bumping and shoving good-naturedly.

While weaving in and out of the crowd, Officer Peak is of the impression he just saw a young man reach into the outside coat pocket of another man walking in the mall. Peak is about 20 feet from where the incident is alleged to have taken place, and with the movement of the people in and out of his sight, he is not positive, but he is almost sure he is right.

Pushing his way rapidly through the crowd, Peak reaches the suspect, takes hold of his arm, and says, "Come with me, young man!"

The young man looks surprised and says, "What for? Who are you?"

Peak responds, "I'm from Mall Security. Come with me."

The suspect pulls back and says, "I will not!"

A struggle ensues, and although Peak is able to take the suspect into custody, Peak suffers a broken nose.

Can the suspect be held civilly or criminally liable for his assault on Officer Peak?

CHAPTER SEVEN

The Use of Force

There are different times when law enforcement officers, peace officers, private security officers, or private citizens may legally use force against others. Each category does not have the same authority. The list below includes those times when the use of force may be initiated in order to accomplish the purpose, but each category of persons must select from the list those which they have the authority to perform.

1. When serving a warrant.
2. When performing an arrest.
3. When performing a search.
4. When performing a seizure.
5. In self-defense.
6. In the defense of others.
7. To recover property or chattels.

As indicated in Chapter 6, police officers, peace officers, private security officers, and private citizens have varying degrees of privilege regarding the use of force.

Although it is important for security officers to know and understand the times when the use of force is permitted, and the amount of force authorized, it is just as important for them to know that *very seldom, if ever,* is it necessary to resort to the use of force. Should officers be confronted with such a situation, it will almost certainly be *nonlethal* force required. More often than not, *use of force will be either in self-defense or in some type of arrest situation.* Because of these possibilities, it is important that officers be aware of the following:

1. They should thoroughly research, review, and know the laws and statutes of the state covering the use of force in such situations.
2. Voluntary cooperation must be requested first before the application of any use of force when only property rights are at stake.
3. In self-defense, use of force is permitted, but only in the amount used by the attacker, *and the arresting person must not be the attacker.* More force than that being used by the attacker is not permitted, nor may the force continue after resistance, or force used by the attacker, is stopped.

GENERALITIES

During the early part of this chapter we will speak in generalities. We will concern ourselves with the general use of force without regard to a specific instance or class of person.

For the most part, rules and conditions for the use of force are the same, or at least very similar, throughout all of the states and jurisdictions.

The authorities for the use of force under the various circumstances by the police, peace officers, private security officers, and private citizens are the same as those authorities defined in the previous chapter on arrest.

USE OF FORCE WHILE MAKING AN ARREST

When making a valid arrest, sufficient and reasonable force may be used to overcome any resistance on the part of the subject being arrested. *When resistance by the subject being arrested ceases, all use of force by the person making the arrest must cease.*

If the arrest is not valid or legal, the use of any force is unreasonable. You will be held strictly and totally responsible.

FELONY ARREST

> The use of any significant force, ...not reasonably necessary to effect an arrest...as where the suspect neither resists nor flees, or where force is used after a suspect's resistance has been overcome, or his flight thwarted... would be constitutionally unreasonable. (*Kidd* v *O'Neil*,774, F 2d 1252, 1256–57, 4th Cir. 1985)

MISDEMEANOR ARREST

> An officer is never justified in using deadly force to arrest for a misdemeanor. (*State* v *Wall,* 286 S.E. 2d 68, N.C. 1982)

The use of force to make an arrest has been a source of concern for the legal community and law enforcement for some time. The old Common Law prohibits an officer from using deadly force to prevent the escape of a misdemeanant, but does allow an officer to use force to prevent the escape of a felon, regardless of the nature of the felony. Some states may still use the old Common Law rule, but more and more states are changing or have changed. States that have changed permit, *in some extreme cases,* the use of *sufficient and reasonable deadly force* to effect an arrest of a subject who has committed a *violent felony, such as murder, and who is in flight from the actual event.* It should be repeated that this would have to be an extreme case, and although it is possible to happen to a security person, it is very unlikely. Mere stealing, or flight from a theft situation, is not sufficient excuse for the use of any amount of deadly force against a suspect.

Webster's defines force as "A physical power or strength exerted against a person or thing": for example, "He used force in opening the door." Or "The use of physical power to overcome or restrain a person; physical coercion; violence": for example, "The police resorted to force to disperse them."

There are two different degrees of the use of force that must be understood.

1. The use of nonlethal force.
2. The use of deadly force.

The use of nonlethal force consists of the use of physical action against another person, the sufficiency of which is not of itself possible to result in death or serious crippling injury.

Deadly force is the use of physical action against another, the sufficiency of which, of itself, is possible to result in death.

The basic supposition of the law concerning the use of force in making an arrest is that force may be used only where absolutely necessary. Whenever a subject offers no resistance, there is no need for the use of force. Should the arresting party use any amount of force under such conditions, such use of force is illegal.

In the greatest majority of cases, arrests are accomplished by the simple use of words or a simple act of touching.

THE REQUIREMENT OF DISCRETION

The use of force is one of the powers that comes with the authority to arrest. *This is an awesome power and must be used with discretion.* This use of force power, although granted, is scrutinized very closely by the courts, and rightfully so. In the excitement of an event, or as a result of resistance on the part of the suspect, tempers can be lost, personal prejudices come to the front, and abuses of this power exceed the accepted norms. Such abuses are dealt with severely by the legal system.

The amount of flagrant violations of this power that have taken place in recent times indicates that perhaps the courts are not yet strict enough. Abuses dealt out to an already incapacitated suspect, or continued upon a totally helpless suspect, should receive the most severe punishment the court is able to give.

The use of force automatically brings a public response if the matter becomes known to the public. Seldom, if ever, does it bring compliments, but it almost always brings complaints. These complaints very often result in charges and lawsuits.

There is no question that certain circumstances require the use of some amount of force, just as certain illnesses require a certain amount of medication. Too high a dose, in either situation, can only be harmful, and may even result in death.

USE OF FORCE WHEN SERVING A WARRANT

Very little will be discussed concerning this subject since security personnel seldom, if ever, will be called upon to serve warrants. However, because this is an instance where the use of force is authorized, we will say:

> The minimum amount of force necessary to serve or issue a warrant may be used by those authorized to serve such a warrant.

In Common Law, by statutes, or by case decisions, the general rule for serving warrants requires notice and identification of authority on the part of the server, plus the purpose for the serving.

If the serving is the place where the person to be served is staying, and admission to that place is refused after verbal notice of authority and purpose is made known, the person or persons executing the warrant may force entry by breaking in.

USE OF FORCE TO EFFECT A SEARCH

A search of a suspect, incident to a lawful arrest of a suspect, is authorized for all levels of arresting parties. It is a matter of safety. The suspect may be searched, along with the immediate surrounding area in which the suspect could reach a weapon or destroy evidence.

If, in order to accomplish this search, restraining force is necessary, it may be used. The amount of force being used must be in direct proportion of the amount of resistance being used, or less. It can never be more. When the resistance stops, the use of force must stop.

EXAMPLE

> An officer is summoned to a peace disturbance in an apartment building on the outskirts of town. Upon the officer's arrival, it is determined the apartment dweller is wanted on an outstanding warrant for narcotics possession. He is arrested at the apartment on the strength of the outstanding warrant and taken to the police station.

After questioning the suspect, the police have reason to believe he may have had narcotics in the room at the apartment where he was arrested.

What is the Next Move of the Police?

Since the officers did not perform a search incident to the arrest while at the scene of the arrest, they lost any right to search that area without a warrant. If they wish to make a search of the area at this time, they must first obtain a search warrant. Any other search would be illegal.

EXAMPLE

Officer Adams recognizes a wanted suspect strolling on the sidewalk just ahead of his police car. He pulls in behind the man, gets out of his car, calls out his identification, and orders the suspect to halt. Officer Adams then directs the suspect to "spread against the side of the patrol car."

When the suspect is leaning against the car, Officer Adams places him in handcuffs and directs him to stand up in order to be searched. As Adams stoops to check the suspect's lower trouser legs, the suspect kicks him, knocks him down, and begins to run.

What is Adams' Next Move?

Adams, not being injured, gets up and goes in foot pursuit after the suspect. When he comes up behind the running man, Adams grabs the suspect by the handcuffs and elevates the man's bound arms up behind his back. This puts the suspect off balance, causing him to fall to the ground. Adams then helps the man to his feet, keeping hold of the cuffs, using them for leverage. Adams uses just enough leverage to keep the suspect under control and escorts him to the patrol car. After securing the suspect in the back seat, Adams calls for a backup car in order to have help in performing the search of the suspect.

Anyone making a legal warrantless arrest has the power, without the benefit of a search warrant, to make an immediate search of the arrested person and the area and things under the immediate control of the suspect. This exception to the rule of search warrants is permitted so as to allow for the seizure of weapons or other items that may be used to attack the person making the arrest, to prevent an attempted escape, or to destroy evidence.

FORCED ENTRY FOR AN ARREST OR A SEIZURE

Even when officers are armed with a search warrant, there are certain rules with which the searching or entering officers must comply.

Before officers may force their way into a dwelling of any kind with intent to arrest a person inside the building, they must follow a procedure. They must knock first, announce their authority and purpose, and then demand entry. *Miller* v *United States,* 357 U.S.301,313–14, 78S.Ct.1190–1198,2L.Ed.1332, 1340–41, 1958)

In essence, what the above decision dictates is that an arresting officer or officers must do something such as the following: Knock or bang on the door and almost instantaneously call out in a loud voice, "*Police—we have a warrant for the arrest of*____________," or, "*We have a search warrant,*" and then demand entry before using any force to break down the door or other means to gain forceful entry.

Failure to knock, announce, and demand admittance will be excuse under the following conditions only:

1. When the officer's purpose is already known to the offender or other person upon whom demand for entry is made.
   Miller v *Supreme Court of the U.S.* 357, U.S. at 310 78 S.Ct. at 1196, 2L.Ed.2d at 1338
2. When the personal safety of the officer or other persons might be imperiled.
   United States v *Guyon,* 717,F 2d.1536,6th Cir. 1983
3. When the delay to knock and announce might defeat the arrest by allowing the offender to escape.
   State v *Fair,* 211 A.2d.359 N.J.1965

4. When knocking and announcing might allow persons inside to destroy evidence, such as in drug cases. *Ker* v *California,* 374, U.S.23,83S.Ct.1623,10L.Ed.2d 726, 1963

USE OF FORCE IN EFFECTING A SEIZURE

For the most part, the rules for the use of force in effecting a seizure are the same as, or very similar to, those indicated in the previous paragraphs on the use of force in effecting a search. This too is a case where it would be wise for the reader to review the statutes of one's own state or jurisdiction. Search and seizure are normally so closely allied that there is seldom any differences in the rules.

USE OF FORCE IN SELF-DEFENSE

This topic is so important and covers so much that the author feels it is important for readers to know from where the research for this topic was taken so readers will be able to research from the same source if they care to do so. Research was taken from the *Restatement of the Law of Torts,* 2nd revised and enlarged edition. The *Restatement of the Law of Torts* is considered by all jurisdictions and states to be something that carries authority and weight, to the degree that in courts throughout the country and even the federal courts, it is used as an authoritative reference.

SELF-DEFENSE BY FORCE NOT THREATENING DEATH OR SERIOUS BODILY HARM

1. A person is privileged to use *reasonable force,* not intended or likely to cause death or serious bodily harm, to defend himself (herself) against unprivileged harmful or offensive contact or other bodily harm which he (or she) *reasonably believes* another is about to inflict on him (or her).
2. Self-defense is permitted under the conditions stated in 1 above although the person correctly or reasonably believes he (she) can avoid the necessity of so defending himself (herself),
   a. by retreating or otherwise giving up a right or a privilege, or
   b. by complying with a command with which he (she) is under no duty to comply or which the other is not privileged to enforce by the means threatened.

SELF-DEFENSE BY FORCE THREATENING DEATH OR SERIOUS BODILY HARM

1. A person is privileged to defend himself (herself) against another by force intended or likely to cause death or serious bodily harm when he (she) reasonably believes that
   a. the other is about to inflict upon him (her) an intentional contact or other bodily harm and that
   b. he (she) is thereby put to peril of death or serious bodily harm or ravishment, which can safely be prevented only by the immediate use of such force.
2. The privileges stated in the above subsection exist although the person correctly or reasonably believes he (she) can safely avoid the necessity of so defending himself (herself) by
   a. retreating if he (she) is attacked within his (her) dwelling place, which is not also the dwelling place of the attacker, or
   b. permitting the other to intrude upon or dispossess him (her) of his (her) dwelling place, or
   c. abandoning an attempt to effect a lawful arrest.
3. The privilege stated in subsection 1 does not exist if the person correctly or reasonably believes that he (she) can with complete safety avoid the necessity of so defending himself (herself) by
   a. retreating if attacked in any place other than his (her) dwelling place, or in a place which is also the dwelling place of the other, or
   b. relinquishing the exercise of any right or privilege other than his (her) privilege to prevent intrusion upon, or dispossession of his (her) dwelling place or to effect a lawful arrest.

The Character and Extent of Force Permissible

1. A person is not privileged to use any means of self-defense which is intended or likely to cause a bodily harm in excess of that which the person correctly or reasonably believes to be necessary for his (her) protection. If the person does exceed the degree of force which he (she) is privileged to use in self-defense under the circumstances, he (she) will be liable to the extent to which he (she) has exceeded the privilege.

Use of Force in Excess of Privilege

1. If a person applies a force to another which is in excess of that which is privileged,
   a. the person is liable for only so much of the force as is excessive,
   b. the other's liability for an invasion of any of the person's interests of personality which the other may have caused is not affected,
   c. the other has the normal privilege to defend himself (herself) against the person's use or attempted use of excessive force.

USE OF FORCE IN DEFENSE OF OTHERS

A person is privileged to defend a third person from a harmful or offensive contact or other invitation of his (her) interests of personality under the same conditions and by the same means as those under and by which he (she) is privileged to defend oneself if he (or she) reasonably believes that,

1. the circumstances are such as to give the third person the privilege of self-defense, and
2. his (her) intervention is necessary for the protection of the third person.

The person acting in defense of a third person does not take the risk that the person for whose protection he (she) interferes is actually privileged to defend himself (herself). The person may be guided by appearances which would lead a reasonable person in his or her position to believe that the third person is so privileged.

USE OF FORCE IN DEFENSE OF PROPERTY

Defense of Possession by Force Not Threatening Death or Serious Bodily Harm

A person is privileged to use reasonable force, not intended or likely to cause death or serious bodily harm to prevent or terminate another's intrusion upon his (her) land or chattels, if

1. the intrusion is not privileged, and
2. the person reasonably believes that the intrusion can be prevented or terminated only by the force used, and
3. the person has first requested the other to desist and the other has disregarded the request, or he (she) reasonably believes that a request will be useless or that substantial harm will be done before the request can be made.

Defense of Possession by Force Threatening Death or Serious Bodily Harm

The intentional infliction upon another of a harmful or offensive contact or other bodily harm by a means which is intended or likely to cause death or serious bodily harm, for the purpose of preventing or terminating the other's intrusion upon the person's possession of land or chattels, is privileged if, *but only if,* the person reasonably believes the intruder, unless expelled or excluded, is likely to cause death or serious bodily harm to the owner or a third person whom the owner is privileged to protect.

In the case of *Katko* v *Briney* (183,N.W.2d 657 Iowa, 1971), the plaintiff sued for damages resulting from serious injury caused by a 20-gauge spring shotgun set by the defendants in a bedroom of an old farm house that had been uninhabited for several years.

The plaintiff and a companion had broken into the old farmhouse to find and steal antique bottles and dated fruit jars. The defendants had rigged a 20-gauge spring shotgun to shoot low, but to shoot anyone entering the north bedroom of the farm house. When the plaintiff and his companion entered the room while breaking and entering with the intent to steal, the shotgun was discharged and the plaintiff was seriously injured in the legs.

In the statement of issues the trial court stated the plaintiff and his companion committed a felony when they broke into the defendants' house. In instruction no. 2, the court referred to the early case history of the use of spring guns and stated under the law that their use was prohibited except to prevent the commission of felonies of violence where human life is in danger. The instruction included the statement that breaking and entering *is not a felony of violence.*

Instruction no. 5 stated, "You are hereby instructed that one may use reasonable force in the protection of his property, but such right is subject to the qualification that one may not use such means of force as will take human life or inflict great bodily injury. Such is the rule even though the injured party is a trespasser and is violation of the law himself."

Instruction no. 6 stated, "An owner of premises is prohibited from willfully or intentionally injuring a trespasser by means of force that either takes life or inflicts great bodily injury; and therefore a person owning a premise is prohibited from setting out 'spring guns' and like dangerous devices which will likely take life or inflict great bodily injury, for the purpose of harming trespassers. The fact that the trespasser may be acting in violation of the law does not change the rule. The only time when such conduct of setting a 'spring gun' and like dangerous device is justified would be when the trespasser was committing a felony of violence or a felony punishable by death, or where the trespasser was endangering human life by his act."

Prosser on Torts, 3rd edition, pages 116–18 states: "The law has always placed a higher value upon human safety than upon mere rights in property; it is the accepted rule that there is no privilege to use any force calculated to cause death or serious bodily injury to repel the threat to land or chattels, unless there is also such a threat to the defendant's personal safety as to justify a self-defense.... Spring guns and other man-killing devices are not justifiable against a mere trespasser, or even a petty thief. They are privileged only against those upon whom the landowner, if he were present in person, would be free to inflict injury of the same kind."

CHAPTER 7 PROBLEMS

PROBLEM 1

Security Officer James Monroe is patrolling on foot in the Meadowlark subdivision. Meadowlark is a new housing development consisting of beautiful new homes in the price range high above the average. The community consists of some very affluent residents, affluent enough to want to provide for their own private security.

It is 1:30 A.M., and Officer Monroe is flashing his spotlight in the spacious area between 901 and 903 Floral Place when the beam falls on a male subject exiting a side door of 903 Floral Place. The subject is carrying a stuffed pillow case.

Drawing his pistol, Officer Monroe calls out to the subject, "Halt! Security officer! Stand where you are!" The suspect remains still for a moment, but as Monroe begins to approach, the suspect drops the pillow case, turns, and begins to run away. Officer Monroe once again calls "Halt!" When the suspect fails to stop, Monroe fires one shot, hitting the suspect in the back of the head, killing him instantly.

Is this a good shoot, and is Officer Monroe justified in the use of force?

PROBLEM 2

Jim Collins is a security officer at the Hamburger Heaven Drive-in. There have been many instances of late in which teenagers have become quite rowdy and have caused some property damage to fixtures and equipment inside the drive-in.

At 10:30 P.M. Monday, 17-year-old Sammy Smith, 5'9", 175 pounds, and his girlfriend, 16-year-old Silvia Adams, 5'3", 115 pounds, are in the drive-in. Sammy is trying to impress his girlfriend with his "tough guy" act. He becomes boisterous and nasty to the waitress, to the point where he is abusive. The waitress, having no defense, goes crying to the manager. The manager calls Officer Collins from the parking lot and informs him that he wants Smith and his girlfriend removed from the premises.

When Collins tells Smith, in a nice way, to leave, Smith remarks, "I don't have to take anything from a rent-a-cop" and takes a swing at Collins. Collins ducks the blow and strikes Smith across the side of the head with his night stick, cutting him severely. While Smith is on the ground, Collins handcuffs him very tightly, and tells the manager to call the police.

When the police arrive, Collins tells them what happened and advises the police he wants to press charges against Smith for disturbing the peace, destruction of property, resisting arrest, and assault.

Was the arrest good? Would it stand up in court?

PROBLEM 3

Art Anderson is a security officer for Safety Security Company. Art recently retired as a floor manager for Smith Dry Goods Company. Art liked his security job of watching the monitors in the lobby of the T.B. Jones Investment Company. The hours were long, but the job was just right for him. He didn't have to be on his feet all night. Being alert and watchful was not a problem, but Art's arthritis acted up when he had to stand any length of time. It was hard for him to move at times, but he hid his pain and kept on because he couldn't get by on his Social Security alone. He retired at age 62, and received only $500 per month.

At about 11:00 P.M. Tuesday, Art noted one of the second-shift mail boys walking through the lobby toward the front door. He carried a package under his arm.

Company rules state that all packages leaving the building had to be checked by security, so Art called to the young man to stop so he could examine the package.

The young man, about 20 years of age, 6' tall, 195 pounds, acted like he was insulted by Art's request and gave him some back talk.

Art tried to calm him down. He explained that the inspection was not his idea but a company rule, and that he would have to submit to inspection the same as anyone else.

With that, the young man pushed Art backward, causing him to fall over a lobby ash tray. Art fell hard, and he began to worry when he saw the young man approaching him while Art was still on the ground. Art, from a sitting position, hit him with a blackjack, knocking him unconscious.

Was the use of force (blackjack) in this case proper or excessive use of force? Explain your position.

PROBLEM 4

Harry Smith and his wife want to go away for three weeks to visit relatives in Canada. However, there has been a great deal of burglary and vandalism in the neighborhood in the last six months. Harry decides to protect his home while he is away and sets up a "spring gun" device with a 12-gauge shotgun pointed directly at the front door of his home.

Should anyone enter Harry's home through the front door, the opening of the door will automatically trigger the device and set off the shotgun.

A burglar, not knowing the device is set up and waiting for him, forces open the door, the gun fires, and the burglar is killed.

Can Harry claim self-defense if charged with homicide?

PROBLEM 5

Speed Johnson was playing softball with some friends one Saturday afternoon when two police officers approached him and said they had an anonymous tip he had held up a liquor store Friday evening. They placed Speed under arrest.

Speed knew there must be a mistake because Friday evening was his usual poker game, which his father and brothers took part in at his father's home.

Realizing he was innocent, Speed refused to go with the police. When Officer Adams went to grab hold of Speed, Speed hit him in the head with the bat he was holding. Officer Jones subdued Speed and he was taken to the station. While at the station, another officer brings in the real robber, who has confessed.

Speed is then booked for assaulting a police officer. Is this a legal or illegal booking?

CHAPTER EIGHT
Search and Seizure

There are times when a private security officer can and does become involved in a search and seizure situation. This does not happen to the extent that it does to a peace officer, but it must be remembered that there are states and jurisdictions in which, under certain conditions, a private security officer has the same authority and responsibility as a police officer within the confines of a client's property.

Thus security officers cannot ignore this fact and say it will never happen to them. Individuals who feel that the laws of search and seizure don't affect them should still study the material if for no other reason than to make them more knowledgeable.

It should also be remembered that if a private citizen makes a civilian arrest that person has the same search right as anyone else in order to protect oneself from hidden weapons.

STOP AND FRISK

Because an incident of "stop and frisk" may be the most probable situation for a private security officer or a private citizen to become involved in, we will review the foundation upon which the authority for stop and frisk is based.

It is necessary to determine if "stop and frisk" can be distinguished from "arrest and search," knowing that justification of stopping and frisking someone is less than probable cause to arrest and search.

The Supreme Court ruling in *Terry* v *Ohio* (392, U.S.1,20 L.Ed.2d 889, 88Sup.Ct.1868, 1968) is probably the most widely known case involving stop and frisk and a case upon which precedents have been set.

In *Terry* v *Ohio,* an officer, while on routine patrol duty, observed Terry and a co-defendant act in an unusual manner in front of a store, which led the officer to believe the men were "casing" the place in preparation for a robbery.

This was pure speculation on the part of the officer. He feared the men were armed.

He approached the men, identified himself as a police officer, and asked questions about their identification and their reasons for being where they were, in front of the store. The suspects mumbled something in answer to the officer's inquiries, which made the officer uneasy. He patted down the outside clothing of suspect Terry and found a gun on his person. This made the officer very wary and he ordered the two men inside the store where he searched the other man and found another gun. The officer immediately placed the men under arrest for "carrying a concealed weapon."

At a pretrial hearing, the defense made a motion to suppress the evidence, stating the search was a violation of the Fourth Amendment guarantee of freedom from searches based on suspicion and without probable cause.

The trial court denied the motion to suppress the evidence, making a distinction between an investigatory stop or detention, and an arrest.

In the appeal to the Supreme Court it was stated that "simple good faith on the part of the arresting officer is not enough."

Mr. Chief Justice Warren, speaking for the Court, said, "It does not follow that because an officer may lawfully arrest a person only when he is apprised of facts sufficient to warrant a belief that a person has committed a crime or is committing a crime, the officer is equally unjustified, absent that kind of evidence, in making any intrusions short of an arrest. Moreover, a perfectly reasonable apprehension of danger may arise long before the officer is possessed of adequate information to justify taking a person into custody for the purpose of prosecuting him for a crime.... We need not develop at length in this case, however, the limitations which the Fourth Amendment places upon a protective seizure and search for weapons. These limitations will have to be developed in the concrete factual circumstances of individual cases.... Each case of this sort will, of course, have to be decided on its own facts."

The stop and frisk doctrine is the result of the reasoning expressed in emergency situations, such as automobile stops, and similar situations when the safety of the officer may be at stake. Naturally, the conduct of the officer during such a stop and frisk will come under the closest kind of scrutiny from a trial court to determine whether any violations of the Fourth Amendment occurred.

While pursuing their duties, officers must be sufficiently skilled so their efforts are not negated because of a violation of the fundamental rights of a suspect or an accused. *It is imperative that the person stopping, frisking, or arresting be fully aware of all the safeguards due the suspect or an accused.*

Remember, in certain jurisdictions and under certain conditions, a security officer is granted the same powers as a sworn police officer, or a peace officer. This being the case, it is important that security officers know the same limitations placed upon them, or the same privileges granted them, as are placed upon or given to a police officer.

Because of the above condition, it is necessary that all security officers become aware of the powers granted them under their license. The range varies from no more power than a private citizen to the same power of a sworn law enforcement officer, within the confines of a client's private property.

This also must be kept in mind: Security officers working in a jurisdiction wherein they have police powers lose those powers the minute they go off duty and leave their designated work area. At that time, unless called in to assist a law enforcement agent in need, the officer is the same as any civilian.

CAUTIONS FOR POLICE AND SECURITY OFFICERS

When an officer falls into the trap of using irregular or questionable practices for the sake of convenience or expediency, he or she is almost certain of making the arrest or seizure useless, regardless of how serious the crime, or who the suspect is.

An officer may not summarily detain an individual and frisk that person for concealed weapons unless the officer has probable cause or reason to believe one is dealing with a dangerous individual. *The officer cannot operate on mere suspicion or hunch* (see *Terry* v *Ohio*). This belief must be justified by circumstances suggesting that the officer's safety, or that of others, is in danger.

As was said previously, but worth repeating, in many jurisdictions the private security officer has the same rights and authority of a police officer, as long as the officer is within the confines of the client's property where he or she is assigned to work. Officers may legally arrest, search, seize, or detain a suspect, or search anything the arrested person is carrying at the time, if the officer has good reason or probable cause to fear for his or her safety. This search should be no more than a "pat down" or "frisk" type search, unless there is an arrest. If an arrest is made, the officer has the right to make an entire body search, provided certain conditions are met.

A possible reason for the officer to go further than an exterior "pat down" or "frisk" search would be if the officer felt an outline of an object, which was obviously a weapon of some kind, through the clothing or in a hand bag or other type of bag.

All such searches should be conducted in a reasonable manner, without intimidation or embarrassment to the person being searched. Whenever possible, the search should be done in public view. However, this is definitely not the case if the search must be a full body search. It a body search is needed, the suspect should be taken to the police station and searched, retaining as much privacy as possible.

Persons conducting body searches should be the same sex as the person being searched, unless there is an emergency situation where life or injury may result from the alternative. There should be at least one unbiased witness to any actual search, if possible.

Weapons, contraband, or stolen property recovered by a security officer during a search should be turned over to the police of the jurisdiction immediately upon arrival of the police to transport the subject to the station. Contraband or evidence seized by a security officer during an illegal search and detention is still admissible in court.

RIGHTS OF THE ACCUSED

The rights of an accused are protected primarily by the Fourth and Fifth Amendments to the U.S. Constitution. The Constitution guarantees to every person accused of a crime the right to a fair and impartial trial by jury. Under this guarantee, the Fourth Amendment protects citizens from unreasonable searches and seizures, and permits them the right to be secure in their homes, papers, and effects, from search as follows:

> The right of the people to be secure in their persons, houses, papers, and effects, against unreasonable searches and seizures, shall not be violated, and no warrants shall issue but upon probable cause, supported by oath or affirmation, and particularly describing the place to be searched, and the person or things to be seized.

The Fourth Amendment constitutes a definite guarantee by the federal government to protect the basic rights of the individual, and is unequivocal in its terms. The fair implications of this Amendment is that no search of one's premises as such is reasonable except under a warrant duly issued, the limits of the search to be fixed, and the scope of the search particularly determined by a disinterested judicial authority. "The right of the people to be secure in their persons, houses, papers, and effects, against unreasonable searches and seizures" relates to the unwarranted intrusion upon the privacy of the individual.

The existence of prohibitions against the government engaging in unreasonable searches and seizures raises the problem of how prohibitions are to be enforced.

ENFORCEMENT METHOD

One enforcement method is to exclude illegally obtained evidence from court trials. The Exclusionary Rule requires that evidence obtained by law enforcement officers using methods which violate a person's constitutional rights is to be excluded from use in criminal prosecution of that person.

The genesis of the Exclusionary Rule started in 1914 with the case of *Weeks* v *United States* (232, U.S.383, 34 S.Ct., 58 L Ed. 652) and was limited to federal law officers.

In 1949, in *Wolf* v *Colorado* (338, U.S.25,27–28, 69 S.Ct.1359, 1361,93L.Ed.1782,1785) the Exclusionary Rule and the Fourth Amendment became applicable to all of the states of the union through the Fourteenth Amendment.

In 1961, enforcement of the Exclusionary Rule was extended to each of the states.

The Exclusionary Rule is a type of preventive medicine. Its basic idea is that if illegally obtained evidence is excluded from trials, the incentive on the part of law enforcement to get this evidence will be removed.

The Exclusionary Rule recognizes that officers and other officials do not deliberately violate constitutional rights of citizens or act with evil intent. Most of the time when these rights are violated it is because law enforcement has a desire to obtain incriminating evidence as effectively and quickly as possible. The government finds it very difficult to fault the desire of officers to do their job in obtaining evidence; therefore, government finds it undesirable to punish the individual officer for his or her efforts by imposing civil or criminal penalties. The removal of the incentive to obtain incriminating evidence illegally is done by excluding such evidence against the suspect in a court of law.

SEARCH AND ITS AUTHORITIES

The definition of a search as used in criminal law is the act of examining the person or property of another person in an attempt to locate certain items of property.

There are three authorities for the search of a person or the property of another:

1. search by consent,
2. search by lawful search warrant provided by the court; and
3. search, incidental to a lawful arrest.

THE CONSENT SEARCH

A *consent search* occurs when a person voluntarily allows a law enforcement officer, a peace officer, a security officer, or even a private citizen to search his or her body, premises, or belongings.

The person giving the consent forfeits any right to object to the search. Any evidence seized during such a search is admissible in court, even though there was no search warrant, and perhaps not even probable cause.

In *Schneckloth* v *Bustamonte,* (412 U.S. 218,228,93, S.Ct.2041,2048, 36, 1 Ed.2d, 845–863,1973) the U.S. Supreme Court ruled thus:

> If the search is conducted and proves fruitless, that in itself may convince the police that an arrest with its possible stigma and embarrassment is unnecessary, or that a far more extensive search pursuant to a warrant is not justified. In short, a search pursuant to consent may result in considerably less inconvenience for the subject of the search, and properly conducted is a constitutionally permissible and wholly legitimate aspect of effective police activity.

The consent search is much faster than going through the procedure of obtaining a warrant and it eliminates the need for establishing the often difficult "probable cause." To protect the Fourth Amendment rights of the persons giving consent, courts examine consent searches very closely. They want to be certain the person giving the consent was not coerced in any manner, by implied threat or covert force, by explicit or implicit means.

If the search by consent of a person or possessions of a person is to be totally effective, it is most important for private security personnel, when involved, to obtain a written statement of consent to search, stating the limits of the search permitted, executed by the subject before at least one witness. The statement should say that the consent is being freely given, and is not induced by threat, duress, or promise of any kind.

> Lack of valid consent was found where officers told the defendant, if he didn't consent, the officers could and would get a search warrant which would allow them to tear the paneling off his walls and ransack his house. (*United States* v *Kampbell,* 574, F.2d.962 8th Cir.1878)

SEARCH BY WARRANT

The second authority to search is by a court-issued *search warrant.*

A search warrant is an order in writing, issued by a proper judicial authority in the name of the people, directed to a law enforcement officer, commanding the officer to search for certain personal property, and commanding the officer to bring that property before the judicial authority named in the warrant.

In these warrants, the court will clearly define the areas permitted to be searched, the hour of the day in which the search is to be conducted, and will frequently, if not most often, specify and limit those items of property that may be sought for by the search.

If private security personnel should, by some rare chance or circumstance, become involved in any way in a warranted search, they must fully understand the limitations as set forth in the warrant, and confine their conduct to these prescribed limitations.

The policy forming the basis for the warrant requirement of the Fourth Amendment was stated by the U.S. Supreme Court in *Johnson* v *United States* (333 U.S.10,13–14, 68 S.Ct.367,369, 92L.Ed.436,440, 1948).

> The point of the Fourth Amendment, which often is not grasped by zealous officers, is not that it denies law enforcement the support of the usual inferences which reasonable men draw from evidence. Its protection

> consists in requiring that those inferences be drawn by a neutral and detached magistrate instead of being judged by the officer engaged in the often competitive enterprise of ferreting out crime. Any assumption that evidence sufficient to support a magistrate's disinterested determination to issue a search warrant will justify the officers in making a search without a warrant. This would reduce the Amendment to nullify and leave the people's homes secure only in the discretion of police officers.... When the right of privacy must reasonably yield to the right of search is, as a rule, to be decided by a judicial officer, not by a policeman or government enforcement agent.

The only legal method that can be employed to search the premises of a private individual is by use of a search warrant.

The function of a search warrant is to authorize that which legally could not have been done without its issuance.

A great difference exists between the right to seize contraband goods and papers under the authority of a search warrant, and the authority to make a seizure under a warrant of arrest. This distinction must be maintained. Seizure may follow as an incident of arrest under warrant, whereas a search warrant is an authorization to search for a discovery. If there is to be an arrest of an offender, it follows the discovery and is viewed to be the result of the search.

The search warrant applies to cases involving either detection or prevention of crime and may not be used for the protection of some private right.

A search warrant is not limited strictly to persons or their homes and property, but may also extend to places of business.

The law governing the use of search warrants allows personal property to be taken under the authority of a properly executed search warrant, from the house or any other place in which it is concealed, or from the possession of any person by whom it was stolen or embezzled, or from a third party in whose possession it may be.

Search and Seizure Incidental to Arrest Without a Warrant

The last authority for search, and the one most experienced by a private security officer, is a *search incidental to a lawful arrest.*

When an arresting officer is making a search of the person, vehicle, or premises of an arrested suspect, a violation by the officer of any of the suspect's rights could negate the search and seizure of anything taken from the suspect.

Search and seizure incidental to an arrest means a lawful, valid arrest, not a trumped-up set of circumstances that are not legal. Determination of whether the arrest and search or seizure are valid and lawful is a matter for the court to decide. The court will review the circumstances and the manner in which the search is made. This will include the reasonableness of the arresting officer's belief that a felony was being committed. It will also include whether, as incident to arrest, articles seized were found in the possession, custody, and control of the person arrested. The court will also decide if articles searched for and seized were being utilized in the commission of the crime for which the arrest was made.

If an officer makes a lawful arrest, a reasonable search and seizure may be made incident to that lawful arrest. This may be accomplished without benefit of a search warrant.

An officer may likewise arrest and take into custody, without need of a warrant, anyone believed by the officer, upon probable cause or reasonable grounds, to have committed a felony. In this case, a search incidental to such an arrest is justified.

The general rule of law is that when a person is legally arrested for an offense, whatever is found on his or her person or in his or her control, and which may be used to prove the offense, may be seized and held as evidence. Any lawfully arrested person may be searched for instruments, or fruits and evidence of a crime. This includes papers and documents of an incriminating nature.

> When a search is based upon a magistrate's rather than a police officer's determination of probable cause, the reviewing courts will accept evidence of a less judicially competent or persuasive character than would have justified an officer in action on his own without a warrant...and will sustain the judicial determination so long as there was a substantial basis for the magistrate to conclude that seizable evidence was probably present.... (*Aguilar* v *Texas,* 378, U.S.108,84 S.Ct.1509, 12L.Ed.2d 723,1964)

Situations often arise where the time and effort needed to obtain a search warrant would negate the enforcement of the laws. It is for this reason certain exceptions have been made and have allowed warrantless searches. (See *Chimel* v *California,* 395,U.S.752,89,S.Ct.2034,23L.Ed.2d,685,1969.)

Prior to the *Chimel* decision a police officer was permitted to search, incident to an arrest, all areas considered to be in the "possession" or "under the control" of the arrested person.

Chimel permits law enforcement to search a person, incident to an arrest, for two purposes:

1. to search for and remove weapons the person being arrested might use to resist arrest or effect escape, and
2. to search for and seize evidence in order to prevent its concealment or destruction.

It is important to note, in the general rule mentioned above, incident to a lawful arrest, a search without a warrant may be made of portable personal effects in the immediate possession of the person being arrested. In addition, the discovery during a search of a totally unrelated object that provides grounds for prosecution of a crime different from that for which the accused was arrested does not render the search invalid.

If the arrest made by a security officer or a private citizen is lawful, the arresting person may generally search the subject for weapons, contraband, and the fruits of the crime.

As in the arrest procedure, sufficient and reasonable force may be applied to perform the search adequately, but not one bit more, and absolutely none if the person does not resist.

In addition to a search of the person, private security personnel making a lawful arrest may also search incidental to that arrest any possessions and property within the immediate proximity and control of the subject at the time of the arrest.

FREQUENT AREAS OF CONCERN FOR PRIVATE SECURITY OFFICERS

Incidental to a lawful arrest, the authority to search beyond the person of the subject has been greatly shortened in recent years. If security personnel effect a lawful arrest within the premises of a retail establishment, and in the lawful search of the subject's person find, say, an automobile key or a key to a nearby locker, *they may not,* without the consent of the subject or a search warrant, extend their search incidental to the arrest, to the automobile or the locker.

An area of concern to private security officers is their right to search the lockers or parcels of employees on the employer's premises. This may be accomplished if certain guidelines are met.

1. It is clearly understood, between the employer and the employee at the time of hiring, that a search or inspection of the employee's parcels, bags, briefcases, etc., may be made at any time by management. This could then be considered an "implied consent" as a condition of employment, and such search would be justified.
2. In this regard, it would be well if the employees were organized under a collective bargaining agreement; such matters would be part of that agreement, or at least the right of such type of search be acknowledged by the appropriate union officials.
3. Conducting this type search, although permissible to be done at random times, could never be used in any manner that would appear discriminatory as to one class or group of employees.
4. The lawful search of lockers, desks, etc., may also be conducted if management clearly declares such items are the property of the management, and are merely being lent to and being used by the employees, and said employees understand this property is subject to search at any time by management.

THINGS TO BE AWARE OF

Generally speaking, while an improper or illegal search by a private person may give rise to civil action, the evidence obtained in such illegal search may be admissible against the subject in a criminal proceeding.

A "stop and frisk" may develop into a seizure or an arrest if both the detention and the search go beyond the allowable purpose of the original stop.

The "frisk or pat down" at the very beginning must be confined to a person's outer clothing. There is no authority, at this stage, to reach into the subject's pockets. However, if during the pat down the officer detects an object that feels like it may be a weapon, the officer may reach inside the clothing or that particular pocket and seize it, if it is a weapon. If the item is not a weapon, but an implement of crime, such as a jimmy, screwdriver, lock puller, etc., the item is admissible in evidence for a crime to which it relates.

Ordinary looking, hearing, smelling, etc., does not constitute a search in a sense that would require a constitutional authority, such as a search warrant. Officers, because of their occupation, are no different from anyone else. If during the tour of duty an officer happens to be where he or she has a right to be, and does not engage in improper conduct, the officer is entitled to observe what is going on around him (her). Looking at things that are in plain sight of everyone does not violate the Fourth Amendment. It is an officer's duty to be alert to everything within sight and hearing of one's location or position, since observations under such conditions may become the basis for probable cause for a future arrest.

Courts have held that merely looking at objects carried on the street, or in parked cars, is not a search; nor has an officer searched if he or she has noticed objects in plain view in a vehicle that he or she has stopped as long as the stopping of vehicle was proper. Once an object is in plain view, a closer examination will not be considered a search. This does not mean, however, that laws can be broken or rights violated in order to get a closer look.

Objects previously abandoned or thrown away may be examined without there being a search constituted. Items openly discarded, such as a bag discarded by a suspect during a chase, may be inspected without constituting a search. There is, however, a question of doubt concerning items discarded within the confines of a suspect's home or property. Items discarded into a waste basket, inside of a dwelling, even though in plain view, should not be searched without a search warrant, unless the basket is moved from the premises. Items discarded into trash containers located in public areas may be examined and seized without violating Fourth Amendment rights.

In addition to one's home, apartments, hotel rooms, and business premises are considered to be protected under the Fourth Amendment. To search them would require a warrant. So also is open land area closely connected with a building, such as a backyard.

Wire tapping and bugging are illegal without a court order.

If in the course of one's duties an officer has legal reason to stop a vehicle and has the vehicle impounded for a legal reason, that vehicle may be searched, including locked areas, by the impounding officer for inventory purposes. Any contraband found during such a search is considered legal evidence and is admissible in court.

Lawful seizures are usually limited to the following:

1. Weapons, consisting of firearms, illegal knives, incendiary devices, explosives, or such instruments as clubs, chains, razors, and other devices employed to inflict harm, pain, or even death on a victim.
2. Instrumentalities of a crime, such as burglar tools, or other mechanical or electrical devices used in burglaries, break-ins, etc.
3. Fruits of a crime such as stolen property in the possession of the person arrested.
4. Bloody clothing or other materials, or clothing matching the description of anyone suspected of a crime; anything that may be evidence and have a bearing on a crime that has been committed.

A frequent dilemma of security personnel occurs after apprehending a subject in a larceny within a retail establishment. A search incidental to the arrest reveals additional property in the possession of the thief. It appears to have been the property of another nearby retailer. Inasmuch as the apprehending officer did not witness the theft of this added property, before any seizure the officer should make certain the title to the property remains vested in the other retailer. The most conservative course to pursue in such a case is, if in doubt, do not seize the added merchandise. Notify the other retailer and permit that individual to interview the subject while still in your custody in order to determine ownership of the material.

Any seizure resulting from an unlawful search cannot be considered as being lawful.

Evidence obtained in connection with a wrongful search by a private citizen is admissible as evidence against the victim of the search, as long as the improper search has not been instigated or is not in concert with law enforcement officers.

CHAPTER 8 PROBLEMS

PROBLEM 1

A security officer assigned to an entry point to a restricted area was confronted by two males carrying briefcases, and two females carrying purses.

Clearly visible was a sign that read, "All persons entering or leaving this installation will have his or her briefcase, packages, boxes, or purses inspected by the security officer."

Upon request of the security officer the two men and one of the females handed their briefcases and purse, respectively, to the officer for examination. However, the second female refused, at which time one of the males turned to her and stated, "You didn't bring that thing with you, did you?" To which the woman replied, "Yes, I forgot about the guard being here. I'll take it outside and arrange it so I can meet you inside."

She turned away and started down the stairs toward the door at the foot of the stairway leading to the building exit. What should the security officer do?

PROBLEM 2

While working the shoplifting detail in a retail store, a private security officer saw a male walking around the side exit of the store. During the time the suspect walked the length of the store, he had placed his hand on his right side (at about the belt line) several times, as if to adjust something. At one point he placed his hand underneath his jacket to make the apparent adjustment. Several yards from the exit door, the man paused and examined a number of items on the display counter. As the man was examining the items the security officer walked past him to get a better look at the bulge he had observed the man adjusting.

The bulge had a semi-circular shape and gave the appearance of a revolver being carried in the man's belt. The officer inconspicuously strolled past the man and proceeded to a point directly between the subject and the side exit door. The officer stopped just inside the door to take a position a few feet to the right of the door. As he turned around to face the interior of the store, the man apparently finished his examination of the items on the display counter and walked toward the exit door and the officer. As the subject moved past the officer and reached for the door, the bulge in the subject's jacket became even clearer. Convinced that what the man carried was a weapon or a stolen item from the store, the officer stopped the subject and inquired as to the nature of the item located beneath his jacket. The man said, "It's none of your business," and again started for the door. What should the security officer do?

PROBLEM 3

There has been an ongoing police investigation into the operation of a large drug ring at the Shady Rest Apartment Complex. Information that the police have developed indicates that not only are drugs being sold there but that the receipts from the deals are kept in the chimney flue of one of the unused apartments.

Not having sufficient evidence or probable cause for a search warrant, the detectives, in desperation, go to the apartment complex owner and tell him their story. They ask his permission to check out the empty apartment where the money is stashed. The owner agrees to permit the search.

1. If police find money or drug evidence in the empty apartment can it be used as evidence in a future trial?
2. If police receive permission from the resident manager, can they search any of the apartments?
3. Can a neighbor next door, who witnessed drug transactions going on, give permission for a search? Would it be legal?
4. If the drug dealers are in the unrented apartment and hear the police coming and decide to run, can the police search them when they finally catch up with the suspects?
5. If while running, a suspect drops a bag of contraband on the ground, and the pursuing police officer witnesses the dropping and stops to pick it up, can the police use this as evidence against the person in a court of law?

6. After catching the suspects and obtaining the contraband, can the police search the empty apartment without the consent of the suspects or without a search warrant?

PROBLEM 4

In the U.S. Supreme Court decision of *Mapp* v *Ohio* (1961) the Court ruled that

A. The Fourth Amendment to the Constitution forbids the use in federal courts of any evidence unlawfully seized by federal officers.
B. The rule excluding evidence unlawfully obtained applies to state courts as well as to federal courts.
C. Suspects in police custody must be advised of their right to counsel.
D. The right to remain silent must be stated to a suspect in police custody.

Which is the correct answer?

PROBLEM 5

Evidence was seized during the search of a suspect. The search was incidental to an arrest without benefit of a warrant. For this evidence to be admissible, what requirements must generally be met?

A. The arrest must have preceded the search, and there must have been probable cause for making the arrest.
B. The arrest may have either preceded or followed the search, but before beginning the search there must have been probable cause for making an arrest.
C. The arrest must have either preceded or followed the search, and if it followed the search it may be justified by what was found during the search.
D. The search has no bearing on admissibility of the evidence; as long as no undue force was used in the search, the evidence is admissible.

CHAPTER NINE
Torts

What is the meaning of this strange word *tort?* What is it? What does it refer to?

> A tort is an act or commission which unlawfully violates a person's right created by law, and for which the appropriate remedy is a common law action for damages by the injured person.

This legal definition, under the heading "Service," is from *Restatement of the Law of Torts,* 3rd edition.

A more common definition might be: "A tort is an act someone does, or fails to do, which violates the rights of another, or others, to which he, or they, are entitled by the laws of the land. The person doing, or failing to do, such an act is subject to being sued, and if the court rules against him, he is required to pay the person, or persons, who have been injured by his action, or failure to act, damages in the form of money payment."

The most important rights protected by the law of torts are the rights of personal security, the rights of privacy, the right to one's reputation, and the rights required in social and business dealings.

A tort is not usually a crime, but a crime is usually a tort.

A crime is an offense against the state or its people, or against the federal government, or its people, and is punished by the state or federal government.

The difference between a crime and a tort is the fact that a tort is *an offense against a person or persons, by another person or persons,* and is redressed or compensated for by making the party, or parties, who commit the tort compensate the party, or parties, whose rights have been violated or infringed upon. Compensation for the harm done is usually in the form of the payment of *damages* (usually money) by the offender to the persons offended.

Included under the heading of torts are miscellaneous civil wrongs ranging from simple, direct interferences with a person, such as assault, battery, and false imprisonment, to wrongs involving property such as trespass or conversion, to various forms of negligence, to disturbance of intangible interests such as a good reputation, or commercial or social advantage.

These wrongs have little in common, and at first glance appear to be entirely unrelated to one another. It is not easy to discover any general principle upon which they may all be based, unless it is injuries to be compensated for, and antisocial behavior is to be discouraged.

TORT LIABILITY

Liability in tort is imposed by law, without the assent or agreement of either party involved.

Tort liabilities are owed to a person or general classes of persons.

EXAMPLE

A person driving an automobile is under a tort obligation of care to people in the path of his or her car. One is not free to single out one person toward whom he or she will be bound and forget all others. The driver has a duty to drive carefully, according to traffic laws of the area in which he or she is driving. Failure to obey, or to drive without normal care and caution with regard to other drivers or pedestrians, will make the driver subject to arrest, fines, and/or lawsuits from which will come the payment of damages.

ELEMENTS NECESSARY FOR A TORT

A tort requires certain elements in order for it to warrant consideration as a tort. These include:

1. *The existence of a legal duty from the defendant to the plaintiff.* It can be a duty of care, a duty to refrain from doing something, a duty to do a certain act by reason of one's occupation (police officers, doctors, nurses, security officers, lifeguards, bus drivers, etc.).
2. *A breach of that duty.* The breach can consist of a failure on the part of the person who has the duty to either perform or not perform the required duty.
3. *Damage as a proximate result.* The violation of the duty must result in damages to one or more persons, to their property, to their ability to perform, to their good name, or other damages.

TYPES OF TORTS

There are two types of torts.

1. The *intentional* tort is a willful act; it is an act that implies an intent or purpose to injure or hurt. It is an act put into motion with the full consent of the perpetrator's will. It is a voluntary act.
2. The *unintentional* or *accidental* tort is not a willful act. It is an act that takes place without intent on the part of the person committing the act. There is no desire on the part of the perpetrator to hurt, injure, or offend anyone.

A person is not generally held accountable for failure to do something, or failure to perform an act. Although a person's failure or refusal to do a certain thing may bring criticism or condemnation from others or society in general because there may appear to be a moral or ethical duty to perform the act, there is not necessarily a tort resulting from the failure to do the act.

> Duty is a question of whether the defendant is under any obligation for the benefit of the particular plaintiff; and in negligence cases the duty is always the same...to conform to the standard of reasonable conduct in the light of the apparent risk. What the defendant must do, or not do, is a question of standard of conduct required to satisfy the duty. The distinction is one of convenience only, and it must be remembered the two are correlative, and one cannot exist without the other (*Prosser and Keeton on Torts,* 5th edition, 1984)

EXAMPLE

If someone in the middle of a lake appears to be drowning, a person standing on the bank and refusing to attempt a rescue does not commit a tort. However, if a person has a duty to act as a result of a contract or an agreement, such as a lifeguard, and fails to act through neglect or omission, that person has committed a tort.

Osterlind v *Hill,* 263 Mass.73, 160 N.E.301, 1928
Yania v *Bigan,* 397 Pa.316, 155 A,2d.343, 1959
Handiboe v *McCarthy,* 114 Ga.App.541,151 S.E.2d, 1966

THE CIVIL TRIAL

The following list contains the major activities during a civil trial:

1. Opening Statement
   The plaintiff's attorney explains to the trier of the facts, which is either the jury or the judge (in the case of a nonjury trial), the evidence he or she intends to present as proof of the allegations in the civil complaint.

Upon completion of the presentation by the plaintiff's attorney, the defendant's attorney may then explain the evidence he or she intends to present in order to deny those allegations. It is also the option of the defense attorney to reserve his or her opening statement until the plaintiff's attorney has completed calling all of the plaintiff's witnesses.

2. Call Witnesses

   Witnesses should be certain of and well acquainted with the contents of their statements. Unless asked to read a passage from some writing or a book, witnesses should not try to answer questions by reading from a notebook or a paper. Witnesses are permitted to testify to additional facts that follow logically from their testimony, but in this additional testimony they should not contradict their original statements or the given facts in the case. It is not only permitted but also recommended that the attorneys on both sides review with witnesses questions they will ask of them prior to the start of trial.

3. Direct Examination

   When the plaintiff's attorney questions the witness called to the stand on behalf of the plaintiff, or the defendant's attorney does the same thing for the defendant's witnesses, it is called "direct examination." The purpose of direct examination is to allow each witness to tell his or her story. The attorney asking the questions is not allowed to lead the witness by asking questions in such a manner that the questions include the answer the attorney wishes the witness to give.

4. Cross Examination

   As soon as the attorney doing the direct examination of a witness is finished with a witness, it then becomes the opportunity of the attorney for the other side to cross examine the same witness. The purpose of cross examination of witnesses is to permit the opposing attorney to damage the witness's testimony in the eyes of the jurors or to discredit the witness. To accomplish this the opposing attorney tries to find errors, lies, or inconsistencies in the witness's testimony.

   The system is a two-stage process. First the plaintiff calls his or her witnesses. After examining them, the defendant's attorney is given the opportunity to cross examine. When this is concluded it becomes the defendant's chance to call his or her witnesses. After each witness is examined by the defendant's lawyer in direct examination, the plaintiff's lawyer gets the opportunity to cross examine these witnesses. (It is well to note, after a witness is involved in direct examination and then cross examination, each side has the opportunity for re-direct and re-cross examination.)

5. Introduction of Evidence

   In the greatest majority of cases, there is usually physical evidence that either side may wish introduced into the trial as either plaintiff's or defendant's evidence or exhibits. Physical evidence are real pieces of evidence such as documents, photographs, guns, maps, drawings, etc. When this material is to be entered into the case as evidence, the attorney making the request,

   1. asks the judge's permission to have the item marked and identified as the plaintiff's or the defendant's exhibit, and to be entered by an identification number assigned to it. Either a clerk or the bailiff is designated to record such evidence and identify it for the entering side. In the majority of cases this is done by the court clerk, but there may be slightly different procedures in the various jurisdictions.
   2. After the item has been properly marked and identified, the opposing attorney is given the opportunity to examine the item. It is at this time, if upon examination, the attorney wishes to object for some reason, he or she may raise objections, along with an explanation for doing so.
   3. The attorney entering the evidence asks the witness to identify the evidence, and upon doing so the attorney moves that the item be entered into evidence. It then becomes part of the trial material.

6. Closing Arguments

   After all of the evidence is presented, witnesses examined and cross examined, and after each side rests its case, the judge then extends to both sides the opportunity to present a closing argument to the jury or to the trier of the facts. In civil cases it is the usual practice for the plaintiff's attorney to make the first closing presentation, in which he or she sums up the evidence and testimony that has been presented. The argument is presented in a manner that gives the best light to the client and to prove to the jury (or judge) that the plaintiff has proven his or her case by a preponderance of the evidence. This is followed by the argument of the defendant's attorney, which favors the side of the defense.

7. Instructions to the Jury

   After the closing arguments are presented, the judge gives the jury its instructions. Sometimes these instructions are long and complex. They instruct the jury on the points of the law regarding the issues at hand. The judge may have instructions as to how witnesses are to be assessed for credibility, and what weight, if any, is to be applied to various pieces of evidence. The instructions will include the rules of law governing the issues in the case and what criteria jury members are to use in their deliberations.

8. Deliberation of the Jury

   After receiving the instructions of the judge, the jury is retired to the jury room where members pick one of their number to act as foreman (forewoman). This is done to assure that order is maintained during the jury's deliberations. When jury members have discussed the evidence and testimony heard in court and have applied and are guided by the judge's instructions, jurors try to reach a verdict. This may take some time, or it may take very little time. When they have reached a verdict, the foreman of the jury notifies the bailiff they have reached a verdict and are returned to the courtroom. The foreman turns the verdict over to the judge, who reads it aloud to the courtroom.

IMPORTANCE OF THE TORT PROCESS

Although adjudication, or trial by judge or jury, is not often resorted to as a means of resolving tort disputes, the custom has an importance to the tort process. The importance goes beyond the few adjudicated cases. It becomes the foundation upon which future cases of similar circumstances may be decided.

OPINIONS AND THEIR PURPOSE

Appellate courts make the largest number of written opinions. Opinions are normally written by trial judges only in the federal system and not by state trial judges.

The opinion of an appellate court is written by jurists who heard the case on appeal. It states the result reached and the reasons for doing so. If more than one judge heard the appeal, a concurring opinion may be written by a judge who concurs with the decision, but stating his or her own reasons for reaching the decision. These may be entirely different from the reasons of the other judges, but the results are the same. A judge who dissents, or does not agree, has the privilege of writing his or her own reasons for disagreement.

Written opinions of the appellate courts serve numerous functions. They explain to the parties in the case why the court decided the way it did. The appellate decision is of added importance in cases ordering new trials. In that instance the opinion serves as a lesson in the law to the judge who will preside at any new trial. It spells out the mistakes made in the first trial, mistakes that made a new trial necessary.

Appellate court opinions are published and are made available to the public. These opinions are sources of law. They are a part of the legal and moral environment influencing how we conduct ourselves in society. Of greater importance is the fact that opinions from prior cases are a source of law used in the resolution of future legal disputes. They become part of our legal system and have precedent value many years after they have been written.

FUNCTIONS OF THE COURT

When a plaintiff brings a tort case before the court, that individual is seeking a legal remedy to the conduct of a defendant who harmed him or her. In deciding whether to grant the requested remedy, the court performs three functions:

1. It determines the relevant facts.
2. It states the applicable rule of law.
3. It applies the rule of law to the facts in order to reach the proper result.

THE FACTS

In finding the facts of a case, the court determines what happened. In a tort case this means conduct of the parties in the case and the circumstances surrounding that conduct. If, after hearing the evidence, the court finds against the defendant, it concludes that the event in question occurred at a certain place and at a specific time.

THE APPLICABLE RULE OF LAW

When the court states the law governing a case, it declares there is a generalized rule calling for certain legal consequences that follow from a particular set of facts. For example, the rule of law may be declared to be thus:

> If while at a hockey game, a person shall intentionally strike another person when the game is over, and if the blow causes harm, then the person who struck the blow shall pay damages to the person struck. *Take particular note—the statement of facts is very general.* These facts are apt to be similar in more than one occasion. The characteristic of generality would not be present if the facts stated in the rule were specific as to the time and the place the blow is struck, the part of the body struck, and the harm it caused.

Note that the facts are connected to the consequence of the act. The connection is an "if-then" proposition. The portion of the rule following the "if" identifies the facts that are relevant. The portion following the "then" gives a description of the legal consequences of the act. Therefore, the *substantive* law of torts consists of rules of conduct that match up legal consequences with generalized descriptions of fact patterns. It states that a certain group of generalized, nonspecific facts put into action by a defendant will result in a certain legalized penalty to the defendant.

APPLICATION OF THE RULE OF LAW

In applying the rule of law to the facts of an individual case, the court compares the law to the facts as it determines them to be, with the general description of the facts in the rule of law. The outcome of the case depends on whether the particular facts fit within this general description.

FUNCTIONS OF A JURY AND A NONJURY TRIAL

When a jury is involved in a trial, the three functions of fact finding, law declaring, and law application are divided between the judge and the jury. The judge declares the law. The jury finds the facts, and in most cases applies the law to the facts. Occasionally times arise when the trial judge may perform the law-applying function instead of the jury. This takes place in the case of a "special verdict." In a special verdict, the judge sends specific questions to the jury concerning the facts in the case. From their answering statements to these questions, the judge applies the law governing the case to the facts so stated. Assigning the function of fact finding to a jury is generally a matter of constitutional command. The Seventh Amendment of the U.S. Constitution provides for the right to a trial by jury in federal courts, and in all "suits at common law, where the value in controversy shall exceed twenty dollars." State constitutions provide, with some variations, for a jury trial in civil cases. The determination of the applicable rule of law is the exclusive function of the judge.

THE EFFECT OF THE DIVISION OF FUNCTIONS

The division of functions between the judge and jury has an effect on what issues may be appealed. As it is improper for a trial judge to invade the province of the jury and decide issues of fact, it is improper for an appellate court to review and set aside decisions of fact by the jury. The constitutional right of a trial by a jury of one's peers would be of little value if an appellate court were free to substitute its own findings of fact for those of the jury, reached in a trial. The only issues an appellate court may decide for itself are issues of law...that is, declarations of the law made either expressly or implied by the trial judge at the trial. The judge's instruction to the jury and the jury's verdict are the means by which the allocation of functions between the judge and jury is accomplished.

THE JUDGE'S INSTRUCTION TO THE JURY

The judge's instruction to the jury contains a statement of the law governing the case at hand. It contains a description of facts, which under law must be found by the jury in order to support recovery by the plaintiff. If the jury finds the facts in the case fit within the pattern described by the rule of law contained in the judge's instructions, the jury should return its verdict in favor of the plaintiff. If the facts do not fit within the rules, the jury's verdict should favor the defendant.

OBJECTIVE OF THE TORT PROCESS

Our legal system is a human institution designed to accomplish human purposes. Law reflects the kind of society in which we wish to live, and the goals we wish to achieve.

One of the obvious objectives of the tort process is to punish a wrongdoer. This objective seems appropriate in classic cases in which liability is imposed for harmful battery; cases in which one person deliberately and without justification punches another in the jaw, or hits him with a baseball bat, or shoots him. In these cases, we may impose tort liability as a fitting punishment. It is an appropriate response by society to those types of wrongdoing.

Another objective of imposing liability may be to deter wrongful conduct. Liability may be imposed in the expectation that it will discourage people ahead of time from behaving wrongfully. Viewed in this manner, the imposition of tort liability is imposed, not as an end in itself, but as a means to accomplish an end. This method of discouraging wrongful conduct might be accomplished in two ways. People may be discouraged from assaulting each other by the threat that they will be forced to pay for the harm they cause, or people who find themselves the victims of aggressive physical conduct may be discouraged from retaliating because of the availability of a tort action against the aggressor.

Statements of policy objective of deterrence assume that rules of tort law are generally known and understood by people in our society, and that people will, most of the time, behave consciously in order to avoid tort liability.

Policies underlying the tort process in terms of punishment and deterrence place the focus of attention on the wrongdoer. By shifting the attention to the wrongdoer's victim, we might conclude that liability is imposed on the wrongdoer so the victim will be compensated in an attempt to make him or her economically whole. The punch, wrongfully thrown, may break another person's jaw. A broken jaw costs money to repair and the victim may be incapacitated. If we picture the victim as a struggling father of six children, suddenly unable to work and confronted with large medical expenses, it is easy to see why one of the policy objectives of imposing tort liability upon the wrongdoer may be to relieve the victim of the medical bills and help him to support his family during his inability to work.

If the object is to help the victim and his family, why do we impose the requirement that the plaintiff must have been injured by a wrongful act? A broken jaw is not less broken, disabling, or expensive when suffered in a fall on a sidewalk.

If the object of tort liability is to help people who need financial help, why not compensate all needy victims of accidents, irrespective of someone's wrongdoing?

The rules governing harmful battery do not limit the right to monetary compensation only to those in financial need. Very wealthy victims of wrongdoing recover damages just as easily. The imposition of tort liability is not limited to persons who can afford to pay, or who have insurance.

FAIRNESS AND JUSTICE

Why punish? Why deter? Why compensate? Let's add another question. What about fairness and justice?

Beyond punishment, deterrence, and compensation, tort liability may be imposed on wrongdoers because widely shared feelings of fairness and justice require us to do so. Stating the policy objectives of the tort process in this manner suggests some of the difficulties encountered in translating social policy into law. Fairness and justice are vague concepts that mean different things to different people. They provide little in the way of specific guidelines for lawmakers. Therefore, those responsible for making basic policy choices and translating those choices into working rules of behavior is the legislature. Legislators base their decisions directly upon their perception of what is fair and just because, unlike judges, they are a representative body politically responsible to their constituents, but they seem to be in a better position than judges to determine which objectives are shared by the people.

Just because a party to a lawsuit states there is an issue of fact to be resolved by the jury does not make it so. The mere filing of an allegation does not make the allegation true. It is up to the party with the burden of providing the existence of fact to come forward with sufficient evidence to justify a reasonable person into believing that fact occurred. Unless a trial judge determines that reasonable persons can disagree about the existence of a fact, a judge will not send the case to a jury. This does not mean the judge will act on his or her own. Courts almost always act in response to requests and demands for action from parties who come before them. The trial judge will take an issue of fact from the jury only when asked to do so by an appropriate motion for a directed verdict.

If the assertion of fact that lacks adequate supporting evidence does not dispose of the issue of liability, the question of liability will be given to the jury, along with binding instructions. As a matter of law, the jury must decide the particular issue of fact for the defendant. A binding instruction is the equivalent of a directed verdict, but it is limited to an issue of fact other than the issue of the defendant's liability.

In reacting to a request for a binding instruction for a directed verdict, courts in most jurisdictions look at the evidence in the light most favorable to the nonaggressive party. But whether the verdict is in favor of the plaintiff or the defendant, the important point is, in rendering a decision about the quality of evidence being sufficient, the trial judge is not finding facts. The judge is making a ruling of law to the effect that a finding of fact by the jury is not necessary.

CONSENT TO A CRIMINAL ACT

A victim's consent to a criminal act will not be a bar to prosecution for the crime. A criminal prosecution is for the public's interest and not that of an individual.

The weight of authority is that consent to a criminal act is not relevant in civil cases; consent does not constitute privilege.

SELF-DEFENSE

Of the nonconsenting privileges, the most important is self-defense. The general rules governing self-defense are set forth in the *Restatement of the Law of Torts.*

Self-Defense by Force Not Threatening Death or Serious Bodily Harm

1. A person is privileged to use reasonable force, not intended or likely to cause death or serious bodily harm, or offensive contact, or other bodily harm, which he reasonably believes that another is about to inflict on him.
2. Self-defense is privileged under conditions stated above, although the person doing the defending correctly and reasonably believes he can avoid the necessity of so defending himself,
   a. by retreating or otherwise giving up a right or a privilege,
   b. or by complying with a command with which the person doing the defending is under no duty to comply, or
   c. enforce by the means threatened.

EXAMPLE

Prell Hotel Corp. v *Antonacci,* Nev.469 P.2d.399,1970

Provocation by threats and insults does not justify apprehension of immediate battery...does not justify self-defense.

Boston v *Muncie,* 204 Okla,603,223,P.2d,300,1951

State v *Woodward,* 58 Idaho, 385, 74, P.2d.92, 1937

Self-Defense by Force Threatening Death or Serious Bodily Harm

1. Subject to the statement in subsection 3 below, a person is privileged to defend himself against another force, intended or likely to cause death or serious bodily harm, when he reasonably believes that,
   a. the other is about to inflict on him or intentional contact or other bodily harm, and
   b. that he is thereby put in peril of death or serious bodily harm or ravishment which can safely be prevented only by the immediate use of force.
2. The privilege stated above in subsection 1 exists although the person using the defense correctly or reasonably believes he can safely avoid the necessity of so defending himself by,
   a. retreating if he is attacked within his dwelling place, which is not also the dwelling place of the other, or
   b. permitting the other to intrude upon or dispossess him from his dwelling place, or
   c. abandoning an attempt to effect a lawful arrest.
3. The privilege stated in subsection 1 above does not exist if the person doing the defending correctly or reasonably believes he can, with complete safety, avoid the necessity of defending himself by,
   a. retreating if attacked in any place other than his dwelling place, or in a place which is also the dwelling place of the other, or
   b. relinquishing the exercise of any rights or privileges, other than his privilege to prevent intrusion upon, or dispossession of his dwelling place, or to effect a lawful arrest.

EXAMPLE

Brasseaux v *Girouard,* La.App.269,So.2d.590, 1972
State v *Johnson,* 261,N.C.727,136,S.E.84,1964
Crawford v *State,* 231,Md.354,190,A.2d.538,1963

Character and Extent of Force Permissible

1. The person acting in self-defense is not privileged to use any means of self-defense which is intended or likely to cause bodily harm...in excess of that which he correctly or reasonably believes to be necessary for his protection.
2. If the person acting in self-defense, in a given case, does exceed the degree of force which he is privileged to use in self-defense, under the circumstances...he will be liable to the extent to which he has exceeded the privilege. The *Restatement of the Law of Torts,* formulates the general rule as follows.

Force in Excess of Privilege

1. If the person acting in self-defense applies force to another which is in excess of that which is privileged,
   a. he is liable for only so much of the force as is excessive,
   b. the other's liability for an invasion of any of the actor's interest or personality which the other may have caused to be affected,
   c. the other has the normal privilege stated in this topic, to defend himself against the actor's use or attempted use of excessive force.

EXAMPLE

Germolus v *Sausser,* 83 Minn.141,85N.W.946,1901

The plaintiff and the defendant had an argument. Plaintiff swung a whip and hit defendant on the arm. The defendant jerked the whip out of the plaintiff's hand and hit the plaintiff with it, inflicting serious injuries. It was held proper to instruct the jury that if it was unnecessary for the defendant to hit the plaintiff for his own self-protection against another battery, the defendant was liable.

EXAMPLE

Monize v *Begaso,* 190 Mass. 87,76 N.E.460,1906

Plaintiff rowed up to the defendant in a boat and threw all of the oars he had at the defendant. The defendant did not row away, which he could easily have done, but jumped into the plaintiff's boat and punched him eight or nine times, as hard as he could. The defendant was liable.

The statements above on self-defense are taken from the *Restatement of the Law of Torts.* They are the opinions of the *Restatement,* which is usually referred to as acceptable reference by courts of all jurisdictions, but acceptance of their interpretation is not mandatory. Students should check the laws and rules of their own state. Each state could have its own interpretation, which may vary from that of the *Restatement.*

DEFENSE OF OTHERS

Although the Common Law privilege to defend third persons against harmful contacts took into account only one's family and household, *it is now generally agreed the privilege also extends to total strangers.*

The force that may be used by someone who intervenes to repel an attack on another is measured by the force the other could lawfully use. If the attack by one person endangers the life of a third person, the person who intervenes to protect that person may use deadly force on behalf of that third person.

The one exception to this general rule arises in cases of a mistake. The person doing the intervening may be mistaken as to which one of two other persons is the aggressor in the conflict, or the person may be mistaken as to the nature of the threat posed by the aggressor. In cases such as these there is a split authority as to whether the privilege of the person doing the intervening is derived from, or independent of, the privilege of the person who benefits from the act.

The prevailing view seems to be the authority as to whether the person who intervenes may use only the force the person on whose behalf he intervenes could generally use. If he is mistaken, even reasonably mistaken, as to the identity of the aggressor, or the severity of the threat, he is liable if he uses what turns out later to be the unwarranted use of too much force.

EXAMPLES

Robinson v *City of Decatur,* 32 Ala.App.654,29 S.O.429, 1947
People v *Young,* 11 N.Y.2d 274,N.Y.2d; 183,N.E.2d,319, 1962

Prosser and Keeton on Torts, 5th edition, is often referred to and used in court as a reference or guide in tort cases. It is considered a valuable reference for the court and very informative. *It is important to note, however, that each state has its own lawmaking authority, and readers should be certain they act in compliance with the laws of their own state.*

The following is an extract from *Prosser and Keeton on Torts,* 5th edition (p. 130, par.2, sec.20):

> All states would now recognize the privilege of anyone to go to the defense of another threatened with any kind of felonious invasion, such as rape or serious bodily harm. It has at times been suggested that the privilege should exist only when there would be a legally or socially recognized duty to intervene.... There is, however, something to be said for the notion that one should "mind their own business," as regards minor dignitary invasions, because intervention can often result in the escalation of a minor disturbance into a serious one, and because the intervenor can often be mistaken about who was the aggressor. It would, however, seem questionable to give any aggressor a cause of action against an intermeddler who acted in behalf of someone who actually had a privilege of self-defense.

DEFENSE OF PROPERTY

When referring to the defense of property rather than the life or welfare of a human being, it is believed that most people would agree there is a significant difference in the nonmonetary value between the two, and, as a result, there would be a lower priority afforded to the privilege to act in defense of one's property.

Once again, referring back to the authority taken from the *Restatement of the Law of Torts,* we will define the general rule concerning the defense of property.

Defense of Possession by Force Not Threatening Death or Serious Bodily Harm

A person is privileged to use reasonable force, not intended or likely to cause death or serious bodily harm to prevent or terminate another's intrusion upon his land or chattels, if

1. the intrusion is not privileged,
2. the person reasonably believes the intrusion can be prevented or terminated only by the force used, and
3. the person has first requested the other to desist, and the other has disregarded the request, or the person reasonably believes a request will be useless, or substantial harm will be done before it can be made.

Defense of Possession by Force Threatening Death or Serious Bodily Harm

The intentional infliction upon another, of a harmful or offensive contact, or other bodily harm, by a means which is intended or likely to cause death or serious bodily harm, for the purpose of preventing or terminating the other's intrusion upon a person's possession of land or chattels, is privileged if, *and only if,* the person reasonably believes the intruder unless expelled or excluded is likely to cause death or serious bodily harm to the person or to a third person whom the person is privileged to protect.

Be very careful in this area. There is nothing that can be taken or stolen that is worth taking a life. A simple intrusion is less than a theft, so unless you are certain the conditions spelled out after "only if" above exist don't even consider the use of any form of deadly force.

FALSE ARREST AND IMPRISONMENT

False arrest or imprisonment is the unlawful restraint by one or more persons, of the physical liberty of another person or persons.

In this particular matter, the words "false" and "unlawful" mean the same thing. False imprisonment or arrest consists of a direct restraint, without legal justification. The main consideration is the lack of lawful authority, and the offense is considered as a trespass to the person imprisoned or detained. Depriving a person of his or her liberty may be done by physical enclosure, by arrest, or by fear of physical restraint other than arrest.

> Imprisonment, although it seems originally to have meant stone walls and iron bars, no longer signifies incarceration. (*Reese* v *Julia Sport Wear,* 260, App.Div.263,21, N.Y.S.2d 99,1940 [locking employee in a store])
>
> Plaintiff may be imprisoned when restrained in the open street. (*Lukas* v *J.C. Penney* Co. 233 Or.345, 378 P.2d 717, 1963 [on an island])
>
> In a traveling automobile.
>
> In a moving train. (*Ward* v *Egan* 64 Can.21,1935)
>
> In an elevator. (*Turney* v *Rodes,* 42Ga.App.104,155,S.E.112,1930)
>
> Confined in an entire city or compelled to go along with the defendant. (*Goodel* v *Tower,* 77 Vt.61,58A,790,1904)

The older idea of confinement has persisted, however, in the requirement that the restraint be a total one, rather than a mere obstruction of the right to go where one pleases. Thus, it is not imprisonment to block plaintiff's passage in one direction only.

One of the major problems in an action for false imprisonment is the nature of the confinement required for liability.

In the case of *Whittaker* v *Sanford,* (110 Me.77,85,A.399,1912) a woman was given complete freedom of movement on the defendant's palatial yacht, but when she went ashore occasionally she was not given liberty to roam where she wished, or to remain there. It was held by the court she had been imprisoned. Judgment for the plaintiff was affirmed.

The following is taken from the court's decision in the *Whittaker* case cited above:

> If one should, without right, turn the key in a door and thereby prevent a person in the room from leaving, it would be the simplest form of unlawful imprisonment. The restraint is physical. The four walls and the locked door are physical impediments to escape. How is it different when one who is in control of a vessel at anchor, within practical rowing distance from the shore, who has agreed that a guest on board shall be free to leave, there being no means to leave except by rowboats, wrongfully refuses the guest the use of a rowboat? The boat is the key. By refusing the boat, he turns the key. The guest is as effectually locked up as if there were walls along the sides of the vessel. The restraint is physical. The impassable sea is the physical barrier.

Whether the area from which the actor prevents a person from going is so large it ceases to be a confinement within the area, and becomes an exclusion from some other area, may depend upon the circumstances of the particular case, and be a matter for the judgment of the court and jury.

Perhaps the most troublesome question is how much coercion must be exercised to make out a tortious confinement? Undoubtedly, the defendant will be held to have confined the plaintiff under circumstances where it remains physically impossible to leave. Consider the court's ruling in the case

> where a gas station attendant drained the water from the radiator of the plaintiff's car and told her she was not to leave until the police came. The court held this to be false imprisonment. (*Cordell* v *Standard Oil Co.,* 131 Kan.221,289,.P.472,1930)

Usually the defendant must intend to confine the plaintiff, there being no liability for negligently caused imprisonments. This position is consistent with that reached in assault and offensive battery, but it stands in sharp contrast to the law of personal injuries, which is dominated by an uneasy mix of negligence and strict liability theories.

False imprisonment has both dignity and compensatory dimensions. A person's dignity is harmed. The question is how much is the compensation to be? When harm is minor, the dignity dominates with compensation. Where major physical harm is suffered by the plaintiff, the principle of negligence is used, which allows for higher compensatory damages.

Must a person know he or she is being subjected to external restraint in order to have cause to sue? In one such case,

> the plaintiff had been placed by his mother at the school kept by the defendant, and it appeared that she applied to take him away. The school master very importantly refused to give him up to the mother, unless she paid an amount which he claimed was due. (*Herring* v *Boyle,* C.M.&R.,137,149 Eng.Rep.1126)

The evidence did not show the plaintiff was subjected to any special supervision, or that he was conscious of restraint. On these facts, the court found he could not maintain an action for false imprisonment.

In order to constitute false imprisonment, the confinement apparently need not be of any particular duration. In this respect, the tort is similar to battery, where the slightest unpermitted contact is actionable or subject to suit.

Does the towing of a plaintiff in an auto, which is being towed because of default in car payments, constitute false imprisonment? In one such case, the court held that it was.

> It is true, the defendant argues, the plaintiff was at liberty to depart, and these employees were not preventing him from doing so, but the result of his departure would have been an automatic parting with the automobile, which he did not desire to part with, and which O'Brien and Baer (the defendant's agents) had no right to take over his protests. While he was in his car, he was in a place he had a legal right to be, and in which case neither O'Brien nor Baer had any legal right to be by force, and when these men hooked the wrecker on and hoisted the front wheels into the air, forcibly dragging the plaintiff down the street in his car, this was unquestionably a restraint imposed on him, and a detention of his person, such as constitutes false imprisonment. (*National Bond & Investment Co.* v *Whithorn,* 276 Ky.204,123,S.W.2d.263,1983)

Can the inability of a plaintiff to leave a ballpark at will be considered false imprisonment? The court says it can (*Talcott* v *National Exhibition Co.,* 144 App.Div.337,128,N.Y.S.1059,1911).

> On October 19, 1908, the plaintiff went to defendant's ball park, in an effort to buy a reserved seat for the game that afternoon, the reserved seats being sold from booths inside the enclosure constituting the park. He was unable to get a ticket because all such seats were sold. When he attempted to leave, he was unable to do so through the ordinary exit gates, because crowds were still pressing through those gates, trying to enter the park and the stands. It appears that the baseball game which was to take place was one of very great importance to those interested in such games, and a vast outpouring of people were attracted to it.

Because of the confusion caused by the crowd, people employed by the defendant would not permit the plaintiff to try to go out through the ordinary entrance and exit gates, and they did not show him any other means of exit. As a result, the plaintiff was detained within the enclosure for about an hour before attendants finally escorted him out through the clubhouse, as an alternative exit. The plaintiff recovered a judgment of $500 for false imprisonment, which was affirmed on appeal, the court saying in part,

> We see no reason to interfere with the verdict of the jury in its finding that the plaintiff's detention was unjustifiable under the circumstances. Damages are awarded in the sum of $500.00. The plaintiff proved no special damage, nor was he obliged to. All damages awarded in cases of false imprisonment partake to some extent of "smart moneys," and the sum awarded him is not so excessive as to justify interference on our part.

Trespass for false imprisonment lies where the plaintiff has been completely deprived of his or her liberty by the defendant. Depriving a person of his or her liberty may be done by physical enclosure, by arrest, or by fear of physical restraint other than arrest.

ASSAULT AND BATTERY

It is sometimes said an imprisonment always includes a battery and an assault. Since an arrest may be effected by submission without touching the plaintiff, and without any physical act putting him or her in apprehension of battery, it is evident the statement is incorrect. In many cases of imprisonment—for example, arrest by touching the plaintiff with words of arrest—it is accurate to say there is an imprisonment, a battery, and an assault.

Personal injury is the term frequently used to describe torts to the body. Usually these are "accidental," but not all. Some are planned. Assault and battery is one of the torts and is one of the oldest torts known to society.

> The interest in freedom from apprehension of a harmful or offensive contact with the person as distinguished from the contact itself is protected by an action for the tort known as assault. No actual contact is necessary, and the plaintiff is protected against a purely mental disturbance of this distinctive kind. (*Kline* v *Kline*, 158, Ind.602,64 N.E.9,1902)

> The establishment of the technical cause of action, even without proof of any harm, entitles the plaintiff to vindication of the legal right by an award of nominal damages. (*Walker* v *L. B. Price Merchantile Co.*, 203,N.C.511,166 S.E.391,1932)

An assault is an open threat of bodily contact with someone, without the person's permission.

Any actual body contact is called "battery."

The difference between assault and battery is that between physical contact and the mere apprehension of contact.

The intent element for battery is identical with that of an assault.

The three most basic elements of intent are

1. a state of mind;
2. about consequences of an act or omission, and not the act itself;
3. it extends not only to having in mind a purpose (or desire) to bring about given consequences but also to having in mind belief (or knowledge) that given consequences are essentially certain to result from the act.

As to the basis of liability for offensive battery, a person is subject to another person for battery if

1. a. the person acts, intending to cause a harmful or offensive contact with the person of the other, or a third person, or to cause an imminent apprehension of such a contact, and
   b. an offensive contact with the person of the other, directly or indirectly, results.
2. a. An act which is not done with the intention stated in subsection (a) does not make the actor liable to the other for a mere offensive contact with the other's person, although the act involves an unreasonable risk of inflicting it, and, therefore, would be negligent or reckless if the risk threatened bodily harm.

In *Richmond* v *Fiske* (160 Mass.34,35 N.E.103),

> A milkman entered the bedroom of the plaintiff customer, shook him to waken him, and then presented him with a milk bill. This was held by the court to be a battery.

The protection afforded against offensive battery does not extend only to cases of direct contact with the plaintiff's person. It also covers contact with "anything so closely attached to the plaintiff's person, that it is customarily regarded as a part thereof, and which is offensive to a reasonable sense of personal dignity."

If the defendant is mistaken as to the identity of the person he or she touches, is that a defense to an action for offensive battery? No.

The essentials of assault are an apparent attempt on the part of the defendant to commit a battery on the plaintiff, thus causing the plaintiff reasonable apprehension of a battery.

The "apparent attempt" is made of the following elements:

a. a physical act,
b. an apparent intent,
c. an apparent present ability to commit a battery.

A physical, bodily act is just as necessary here as in the law of battery. Though an assault is sometimes briefly defined as a threatened battery, a mere threat to commit battery is not enough. Any physical act that seems likely to result in a battery is sufficient.

> In *Gelhaus* v *Eastern Airlines Inc.*, 194,Fed.2d.774, the defendant, an employee of Eastern, exclaimed loudly to the plaintiff, in the absence of other employees, "If you are not off the place by 5 o'clock, I am going to have the guards throw your belongings out in the middle of 36th Street." ...The court held that words and threats, unaccompanied by any attempt to use physical force, did not amount to an assault.

While threats alone do not constitute an assault, what the defendant says to the plaintiff, at the time of the act, may be important in determining whether the apparent intent was present.

> In *United States* v *Richardson,* 5 Cranch 348, the defendant, raising a club over the head of the prosecuting witness, said to her that if she said a word he would strike her.... This was said to be an assault.

It would have been an assault had the defendant said nothing, and words used here could not place the defendant in any better position because of the imposed condition that he had no right to impose.

Since assault, as distinguished from battery, is essentially a mental, rather than a physical, invasion, it follows that the damages recoverable for it are those from the plaintiff's mental disturbance, including fright, humiliation, and the like, as well as any physical illness that may result from them.

DEFAMATION

Defamation is the most complex of all torts. The sources of the complexity lie in the historical origins of the law, and in the difficulties of harmonizing the important but often conflicting societal interests in protection of reputation and in freedom of speech.

In 1964 the U.S. Supreme Court began the process of restriking the Common Law balance between protection of reputation and freedom of speech by holding in *New York Times* v *Sullivan* that defamation of public officials is speech protected by the First Amendment of the Constitution. Later cases have expanded the reach of the First Amendment into the law of defamation, and how much remains of the traditional law is somewhat in doubt.

The right protected by law of defamation is the right of reputation. It is a nonphysical tort.

The general rule is that defamation will not be enjoined unless it is incident to some other tort. See *Kwass* v *Kersey,* (139 W.Va.497,81 S.E.2d,1954) and *Krebiozen Research Foundation* v *Beacon Press,* (334 Mass.86,134,N.E.2d.1,1956).

A statement or other communication to the mind of another is defamatory of a person if it

1. holds him up to hatred, contempt, disgrace, or ridicule,
2. tends to injure him in his office, business, trade, or profession.

The basic elements of a claim based on defamation are:

1. *A defamatory statement*—Not all insults are actionable; the general rule is that to be defamatory, a statement must hold the plaintiff up to "hatred, ridicule, or contempt."
2. *Publication*—The basis of the plaintiff's cause of action is the harm he or she suffers from the reaction of others, and not just hurt feelings.
3. *Harm*—The main categories of damages which may be recoverable are special, general, and punitive. The first two have no exact counterpart in other areas of tort law. What can be recovered depends not only on the kind of harm suffered, but also on what is said and how it is said. Of vital importance to the issue of damages is whether the defamation is oral (slander) or written (libel).

For some types of defamation the plaintiff must prove he or she has suffered special damages to recover at all. The principal distinction is between slander (oral defamation) and libel (written defamation). Subject to important exceptions, there can be no recovery for slander, absent proof of special damages, whereas no such proof is necessary to recover in libel. However, the historical justification for this distinction has come under strong fire.

SLANDER

There can be no recovery for slander, absent special damages, unless what the defendant has said falls into the category of slander per se, in which event there may be recovery without proof of special damages. The theory behind making some statements actionable per se is that some charges are so serious that harm to the plaintiff's reputation resulting in economic loss is almost certain to follow.

THE FOUR CATEGORIES OF SLANDER

1. *Statements that the plaintiff has committed a crime.* In general, not any crime will do. American courts usually take the position that the crime must be serious, one involving moral turpitude.
2. *Statements the plaintiff has a loathsome disease.* There are not many diseases that would lead people to regard the plaintiff with sufficient revulsion to justify the presumption of economic loss. There may be a question if in these times, reference to the HIV virus or AIDS would qualify. Sometime in the future it may be tested in the courts.
3. *Statements damaging to one's business, trade, or profession.* The presumption that the plaintiff has suffered economic loss as a direct result of the defamation has its greatest justification in this category. A person's reputation is important to business, and loss of business is likely to follow when that reputation is disparaged. The defamatory statement must relate to the plaintiff's qualifications to conduct his or her business, or to the way he or she conducts it.
4. *Statements made that a woman is unchaste.* This category of slander per se was late in developing, having been created by statute in England in 1891. Many states have followed the English lead, sometimes by statute, othertimes by judicial decision. Men are not protected by this rule, perhaps because male sexual prowess had traditionally been an admired quality.

LIBEL

The rule that proof of special damage is not needed to support recovery in libel is modified in some states by the libel per quod-libel per se distinction. In the states distinguishing between the two kinds of libel, only libel per se is actionable without special damages. To constitute libel per se, the statement must be defamatory on its face. This is not to be confused with slander in per se, which refers to the substance of the statement. If there is need to look outside the statement for facts which will make it defamatory, it is libel per quod, and the plaintiff must show special damage to recover. An exception to the libel per quod rule is that if the statement would have been slander per se if spoken, proof of special damage is not required.

Libel was extended to include pictures, signs, statues, motion pictures, and even conduct carrying a defamatory imputation, such as hanging the plaintiff in effigy, erecting a gallows before his door, and dishonoring his valid check drawn upon the defendant's bank.

TRESPASS UPON REAL PROPERTY

For centuries the law has furnished substantial legal protection to the owners and possessors of land from interference with the possession, enjoyment, and use of their land. The main bodies of law affording this protection are those governing liability from trespass to land, and for nuisance.

Most people associate the word "trespass" with a deliberate, clandestine intrusion upon another's land by someone "up to no good." Trespassers tend to be thought of as fence-breaking, chicken-stealing no-accounts. However, the legal concept of trespass is much broader, more mechanical, and largely devoid of moralistic overtones.

In Common Law, an action for trespass would lie with any unauthorized entry, either by person or by thing, upon another's land, directly resulting from an intentional tort. The legal interest of the plaintiff protected by such an action was the plaintiff's interest in the exclusive possession of the land in question. If the defendant were bodily picked up and thrown upon the plaintiff's land against his will, no action for trespass against him would hold up because the defendant's entry did not result from a voluntary act by him. But it was no defense that the defendant had stumbled and fallen upon the land, or he had entered the land mistakenly, believing the entry was authorized, or that no such entry had occurred. It would have been a trespass even if the defendant had entered the plaintiff's land in reasonable response to physical threats from a third person. Having committed a trespass, the defendant was liable for all harm resulting from his conduct.

Trespass in early Common Law had very little, if anything, to do with moral culpability. To the contrary. It was one of the earliest and stringent forms of strict liability. The main reason for the strictness lay in the fact that an action in trespass was an important legal means by which a lawful possessor of land could maintain the integrity of his possessory interest. Even in the absence of substantial harm to the plaintiff, or his property, the law allowed the plaintiff to establish legally his right to exclude others from his land. Thus, even the innocently motivated but unauthorized entrant would be liable for at least nominal damages, and for compensatory damages if action harm resulted from the trespass. See *Brame* v *Clark* (148 N.C.364,62 S.E.418, 1908).

In the fourteenth, fifteenth, and sixteenth centuries, English courts developed an important distinction between the entries of another's property resulting directly, or by intention, and those entries resulting indirectly, or accidentally, from the defendant's conduct. For those entries made intentionally, a tort of trespass would exist and be substantial; for entries made without intent or accident, an action known as "trespass in case" existed and was substantial. The major difference between the two forms of trespass was the requirement in connection with the "trespass in case." In this instance, the plaintiff had to show the entry by the defendant was committed either intentionally or negligently, and the entry caused actual harm.

The difference mentioned above concerning the directness or indirectness of the entry or trespass has largely disappeared today, and the actual causes of action as they were known then have been established. However, the impact of these causes as earlier established seems to have survived in the important distinction drawn by the courts today between intrusions on land that are intentional and those which are unintentional.

With respect to intentional intrusions on property, much of the strictness remains. One who intentionally enters another's lands or causes a thing or third person to enter the lands of another is liable to an action in trespass. It makes no difference whether or not the defendant caused actual harm, or whether or not the entry was as a result of a mistake on the defendant's part, no matter how reasonable the mistake may have been, unless the entry was induced by some action on the part of the plaintiff.

Rinzler v *Folsom,* 209 G.549,74,S.E.2d 661,1955
Waschak v *Moffat,* 379 Pa.441,109A.2d 310,1954

With respect to unintentional intrusions, one who unintentionally enters the property of another is liable only if he is reckless, negligent, or is engaging in an extra-hazardous activity. In addition, an unintentional intruder is liable only for harm actually caused by such entry. Thus, the modern system of liability for unintentional entries has developed into a flexible system of fault-based liability for actual harm done to the plaintiff's property.

Chartrand v *New York,* App.div.362 N.Y.S.2d,237,1974

Rules governing liability for intentional trespass on another's property still retain some of the severity from earlier times. However, trespass law has become more flexible by the development and expansion of nonconsensual privileges (entries not requiring consent of the owner) which, in a variety of circumstances, excuse intentional entries upon the land of another person. The *Restatement of the Law Torts,* recognizes no fewer than 21 separate privileges of this sort.

1. The use of premises of public utility.
2. The use of public highway.
3. The use of navigable stream.
4. Travel through air space.
5. Deviation from a public highway.
6. Public necessity.
7. Private necessity.
8. Entry to reclaim goods on land without wrong by the actor.
9. Entry on another's land to relieve the actor's land of goods on the actor's land without the actor's wrong.
10. Entry to reclaim or to relieve land of goods which are where they are by no fault of the actor.
11. Entry on another's land to relieve actor's land of goods on actor's land without his wrong.
12. Entry to reclaim or to relieve land of goods which are where they are by wrong of the actor.
13. Entry to abate private nuisance.
14. Abatement of public nuisance by public official.
15. Abatement of public nuisance by private citizen.
16. Entry to arrest for a criminal offense.
17. Entry to recapture or to prevent a crime and in related situations.
18. Entry to assist in making arrest or other apprehension.
19. Entry to execute civil process against occupant of the land.

20. Entry pursuant to order of the court.
21. Entry pursuant to legislative duty or authority.

PUBLIC NECESSITY

One is privileged to enter land in the possession of another if it is, or if the actor reasonably believes it to be necessary, for the purpose of averting an imminent public disaster. (Taken from the *Restatement of Torts.*)

1. One is privileged to enter or remain on land in possession of another if it is, or reasonably appears to be, necessary to prevent serious harm to
   a. the actor, or his land or chattels, or
   b. the other or a third person, or the land or chattels of either, unless the actor knows, or has reason to know, the one for whose benefit he enters is unwilling that he shall take such action.
2. Where the entry is for the benefit of the actor or a third person, he is subject to liability for any harm done in the exercise of the privilege stated in subsection 1 to any legally protected interest of the possessor in the land or connected with it, except if the threat to harm is made to keep one from entering is made in a tortious manner such as threatening use of deadly force.

SUMMARY OF THE UNIQUE CHARACTERISTICS OF TRESPASS

1. The interest which is sought to be protected by a trespass action is the plaintiff's interest in the exclusive possession of the land.
2. To constitute a trespass, the defendant, or the alleged trespasser, must accomplish an entry on the plaintiff's land by means of some physical tangible agency. The entry must be unauthorized and
   a. intended by the defendant, and
   b. caused by the defendant's recklessness or negligence, or
   c. the result of the defendant's carrying on an ultra-hazardous activity.
3. The circumstances in which a defendant may be privileged to commit an unauthorized trespass are carefully limited by judicial decisions, and there exists no broadly based privilege to enter the land of another, simply because on balance the social benefit of doing so appears to outweigh the risks of harm to the land.
4. Once the defendant is found to have committed an intentional trespass, in the absence of circumstances giving rise to a privilege, the plaintiff is entitled to at least nominal damages, and to injunctive relief if further acts by the defendant threaten similar entries upon his land.

CHAPTER 9 PROBLEMS

PROBLEM 1

Substantive law is that branch of the law concerned with

A. Practice and procedure
B. Rights and duties
C. Immunity from torts
D. Oral contracts

PROBLEM 2

How can assault, which is considered a major or minor criminal act, depending upon the degree and circumstances, be considered a tort?

PROBLEM 3

What constitutes "errors and omissions" and why are they considered as tortious acts?

CHAPTER TEN

Major Control Procedures

This chapter contains a general outline of the basics of some of the important control procedures for security officers in major situations. A special burden is placed on security officers. In order to do their job properly they must know these basics and understand the reasoning behind them.

Security officers following these basics will provide a great deal of assistance to the public and to law enforcement. Their efforts will normally bring a grateful "thanks" from the public and from law enforcement. There is another side to the coin, however. Failure to know and follow these basics in an emergency situation will bring discredit to the officer and the agency which employs him or her.

Material in this chapter is very important to security officers. It is possible that lives may depend on one's conduct under pressure and stress.

Like the majority of people, the police appreciate real help. They find their task very difficult under trying circumstances, and assistance given them in the more serious matters receives their gratitude. But here again, there is another side to the coin. When police are operating under stress, interference, improper handling, or ineptitude will trigger anger and resentment. Any one of these can make a bad situation snowball into a serious or critical situation. When this happens, the law enforcement community wishes it never heard of security officers. At times like this, police lump all security personnel into a single classification—inept. The qualified, dedicated security officer's reputation is harmed along with the person who botches the job. It is difficult for the police to erase the memory of the unprofessional handling of a serious emergency situation.

Law enforcement is serious business. It is a field in which anything less than professionalism is not acceptable and where there is seldom an opportunity to correct a mistake. There is no room for glory seekers, make-believe artists, or hero makers. Those in the security profession who are serious about their work and do their jobs with dedication and pride are always welcome by law enforcement. Those who think security is merely a way to pick up some extra money working a second job wear out their welcome in a hurry.

FELONY SITUATIONS

Security personnel are not often confronted with felony situations, but when and if they are, lack of knowledge of how to handle the situation can be a disaster. Security officers must know and realize the importance of the felony situation and be prepared to do their part correctly. The attitude of "It'll never happen to me" is totally unacceptable. If this is a security officer's attitude, he or she should get out before someone gets hurt or killed. If people are not willing to realize the importance of what they are learning, there is no place for them in the security profession. They may not have the prestige of a police officer, and they may not receive the same pay, but they should know that before going into the profession. These facts should never be used as an excuse for anyone to give less than 100 percent effort.

Assume that in the course of a tour of duty a security officer comes upon what appears to be a felony crime committed within the confines of the client's property. *The first moment of discovery can be the most critical time of the entire incident.* That moment can determine whether an investigation gets off to a proper start, or if it is a failure from the beginning. *This critical moment can belong to the security officer who discovers the crime.*

When security officers first determine there is the possibility of a felonious situation at hand, they become the stand-in for "the first officer on the scene." This designation carries with it grave responsibilities, and they all rest on the security officer's back. If he or she knows the job and how to handle it, this is an opportunity to do both oneself and the security profession proud. It is also the time this same officer can turn the crime into an even greater disaster by not knowing what to do, or by doing it improperly.

The security officer's first actions are of the greatest importance. Never let this fact skip your mind. You never know when you will be called on to take these first actions. If you are called upon, you should proceed as follows:

1. Notify the police immediately.
2. In the case of a serious crime (a felony) make no effort to investigate. Do not enter the crime scene area unless there is an injured party who needs immediate medical attention (item 4 below).
3. Seal off the area as best you can. Let no one into the crime scene area other than the first police officer to arrive. This is the person who should take charge and assume the responsibility from you.
4. Do not touch anything or walk into the sealed-off area, if possible. (If there is a victim in the sealed area who could be alive or in need of medical attention, proceed into the area as carefully as possible to determine the condition of the person.) Be sure to avoid stepping on or kicking anything on the floor that may be considered to be evidence, and remember your entrance path as best you can. If the person is injured, render immediate care as best you can and call for medical assistance. Upon the arrival of medical personnel, instruct them of the seriousness of the situation, and the care they must take not to disturb evidence. Keep them confined to the smallest operating area possible, and be sure they exit on the same path they entered. If there is danger of fire or explosion, *do not wait. Remove the injured person by any means possible to a safe area.*
5. Other than medical personnel, *allow no one to enter the area or to touch anything.* This means the client, or anyone else, except the first police officer on the scene who takes over responsibility from you.
6. Obtain the names and addresses of any possible witnesses, along with their statements. This should be done before they leave the scene.
7. Write out a detailed description of any suspects who were seen, and the details of any actions taken by the suspects, if these acts were witnessed.
8. Upon the arrival of the police, identify yourself, advise them of the steps you have taken, and offer to assist in any manner they require. *At the moment of their arrival, the incident becomes an official police matter and your official function in respect to the matter ceases.*
9. Hold yourself available to answer any questions the police may have of you.
10. Furnish a detailed report of the incident as you know it, together with your actions, in writing, to the police. Furnish a copy to your employer, and to the employer's client. Be sure to keep a copy for yourself.

STRIKE OR RIOT SITUATIONS

There are two similar situations requiring proper handling and procedure on the part of the security officer. These situations are *strikes* on the client's property, and *riots* or disturbances in or near their client's property.

These two situations are similar in some respects, but they also differ greatly in others. Each situation presents its own problems, and these problems become those of the security officer on duty at the time. These problems will become smaller and less complex, however, if security officers use a common-sense approach. They need some idea ahead of time of what to expect. Displays of authority, egotism, brashness, or sarcasm on the part of security personnel have no place. Attitudes or dispositions such as these will definitely antagonize the participants in either situation. Negative attitudes may actually place the security officer, and others, in serious danger. During tense or volatile times, unity and cooperation among client, the security force, the police, and others are essential to a successful conclusion. To attempt to go it alone and be the "know-it-all" only invites disaster.

THE STRIKE SITUATION

Consider the basic procedural policies required of a security officer in the event of a strike on the client's property.

1. *Keep a cool head.* You may be verbally abused, but words or names have never been known to draw blood or break bones. To let slanders, taunts, or innuendoes of people on the picket line unnerve you is to play directly into their hands. *You must remain above all of this, indicating to all you are in control of yourself and the situation.*
2. *Take no sides in a strike issue.* You are there to protect lives and the property of the client, not to become involved in any strike issues. *You must keep your opinion to yourself.* To indicate, or to make known, you have a personal view in a strike issue is to destroy your required position of neutrality as a security officer.
3. *Be firm but courteous.* In all of your dealings with members of both sides in a strike, be firm but courteous. It is not your job to make friends or win popularity contests. To give indication of friendliness to one side more than the other will remove all appearance of neutrality. Loss of a neutral position will mean a loss of effectiveness and possibly control of the situation.
4. *Ignorance is no excuse for serious errors in judgment.* Make every effort to learn the legal rights of the strikers concerning picketing and other strike-related privileges. To make an effort to enforce an issue about which you know nothing is to invite trouble with the strikers, and possibly with the police. It is not very effective to try to offer an apology or make amends after a mistake has been made from ignorance of the law and the people's rights. Make certain any information you have concerning rights and privileges of strikers is authentic and up to date, such as court orders and related matters. If someone makes a verbal statement be sure to double-check for authenticity before taking any action.
5. *Learn who the strike leaders and the picket leaders are if possible.* With this information you will know with whom to enter discussions should it become necessary. If contact with either side is necessary, you must know who is in authority and who can speak for the side with whom you are talking. To speak to others can trigger misunderstandings and cause harm.
6. *Try to keep informed of progress in negotiations,* and of the actions of the leaders of both sides. Being aware of the status and progress in negotiations will provide you with sufficient advance information to aid you in decision making if you must take some action.
7. *Use force only as a last resort.* Force should only be used if violence occurs and you are compelled to defend yourself or others, or the property of your client. You must use every means possible to avoid the use of force, but take a firm, nonviolent approach in all of your actions or dealings. If you believe violence is imminent, contact the police immediately and make every attempt to keep things under control until they arrive. Let them handle the situation upon arrival and stay out of it unless your assistance is required by the police. In such a case, take your orders and instructions from them.
8. *Have a plan of action.* Know ahead of time what you will do in the event of any possible development. Go over in your mind as many possibilities as you can think of. Develop a plan of action for each. Do not wait until something happens before you take action. There can always be the unexpected, but you must be prepared for as many types of incidents as possible. Stick as close to your plan of action as is feasible. Stay coordinated and informed with other security personnel. Make all other officers aware of the plan and of the part they are to play. Remember, in unity there is strength.
9. *Do not make statements or threats.* You are to remain neutral throughout the strike. Statements can be interpreted, under stressful conditions, in such a way as to make it seem you are not neutral. The same is true of threats. Threats just antagonize those who are threatened, and your actions can result in an outbreak of violence or destruction. You must keep in mind that there is always a possibility the people on the picket line may be or have been drinking or taking drugs. Their dispositions are usually not the best to start with, since they are losing money being on strikes, and it only takes a small incident to ignite an explosion.
10. *Call the police immediately when law violations occur or when peaceful activities of nonstrikers or others are threatened.* You can tell when the mood is getting grim. Do not wait for the last second before making the call.

THE RIOT SITUATION

Consider the following basic procedures for a security officer in a riot situation. You will note there are many similarities in these procedures to those of the strike situation. However, in the riot situation there is usually a more volatile condition at hand. The situation requires much more self-control on the part of the security officer. *Be on the alert and do not get yourself drawn into a situation where your actions can be the cause of a situation getting out of hand.*

1. *Keep a cool head.* The same requirements are necessary here as were required in the case of a strike, but to a greater degree. Riots are always more explosive than a strike. All common sense and control on the part of the individuals in a riot situation is lost. Rioters are like a keg of gun powder waiting for someone to light the fuse.
2. *Stay neutral.* Usually neutrality does not come into play in a riot situation, but if it is an issue, remember to abide by it to the utmost.
3. *Don't let your action be the trigger that sets off an explosive situation.* Keep in mind that riots are the result of agitation on the part of a few who resort to lawlessness. They stir rioters to a pitch wherein violence is always a split second away. Many times riot organizers are hoping and waiting for someone to set off an incident behind which they can get a mob to unite. If the rioters are not unified they are much less dangerous than if they become an organized force.
4. *Organization is the key.* Organization is the main tool security personnel have at their disposal against the disorganized rioters. The saying "in unity there is strength" was never truer than in a riot situation. If security forces are trained to do things in unison, they will usually be able to keep things under control. If the rioters can be kept in a state of disunity, and if they can be prevented from forming behind one leader, they will have little if any force behind their movement. Keep in mind, *any action taken by the security officers must be in accordance with the law.*
5. *Act in unison whenever possible.*
6. *Determine who are the leaders.* Pick out the leaders or the agitators. Note their description so later you are able to identify them positively and place them at the scene. Pay close attention to their actions, whether or not they carry or use weapons, and if so, against whom. Note those who make threats, whom they threaten, and if any of their threats are carried out.
7. *Take names and addresses of witnesses.* If there are witnesses present, take their names, addresses, and determine what acts they witnessed. This information will be of great use to the police after the situation is in hand.
8. *Call the police at the first sign of unrest.* Do not wait until things are in a state of turmoil before calling the police. Remember, the police have much more authority to deal with this type of situation.
9. *Place yourself at the disposal of the police.* When police arrive on the scene, place yourself at their disposal and follow their instructions.
10. *Make no threats* and give instructions in a clear and simple manner. Carry out whatever you intend to do. *Do not make statements you don't intend to back up with action. Do not back down once you have committed yourself to action (unless you know the action to be unlawful).* If you show weakness, your problems will only multiply. Making any kind of statement under riot conditions takes much thought and care. Loose talk or threats can be your undoing.
11. *At no time are firearms to be used as a threat or are warning shots fired in an attempt to frighten.* Such actions will only incite members of the mob to violence. Remember, they have usually been brought to a frenzy, and it takes only a small incident to push them over the edge into the use of violence in place of words and noise. *A firearm may only be used as an absolute last resort, in self-defense or the protection of the life of another. Even then, great restraint must be used.* You can repair a broken nose, but you cannot give back the life you have taken. *You are not God, nor are you judge and jury. These facts cannot be stressed strongly enough.*

NATURAL DISASTER SITUATIONS

Fires, earthquakes, and tornadoes are natural disasters that most hospitals, institutions, large companies, and schools have prearranged disaster plans to be put in place in case of need. Actually, many plans are similar, their major difference being the size of the operation owing to the size of the facility involved.

The following are the major considerations that must be taken into account by a security officer.

1. Before accepting assignment in any such facility, security personnel should be made fully aware of the established procedures of the facility and how they personally fit into the plan. If officers are designated specific assignments in the plan, they must know what the assignment is and whether they are to use special equipment in carrying out their assignment. They must know where the equipment is located and its availability. Where security personnel fit into the program and to whom they report for direction are the most important things to be aware of.
2. Keep calm. Do not exhibit fear or panic. You must remain in the presence of those you are assisting. Remember, you may be dealing with the injured, the sick, the blind, those who are bedridden, the elderly, and possibly people who are not mentally capable of thinking for themselves. To exhibit fear or panic to these people, no matter how you feel on the inside, can make a horrible situation even worse.
3. Put your part of the predetermined plan into operation at once, whatever it may be.
4. Once you have completed your assignment, remember that there are still others who will need help. Your assignment could include
    a. leading the blind to safety,
    b. removal of bedridden patients, the elderly, and the injured,
    c. continued use of in-house fire equipment until the professionals arrive,

d. emergency first aid treatment,
e. rescue efforts for those in collapsed areas,
f. comforting of the bewildered and confused,
g. protection from looting,
h. aid to ambulance crews.

It is extremely important and vital to success in any type of emergency situation for security officers to remain calm no matter how much they are churning inside. They must be able to give the victims, families of the victims, and those assisting in any rescue operation or help situation the feeling they are competent, careful, sensible, and knowledgeable about the handling of the situation at hand. Officers must be tactful and be able to provide a feeling of assurance. To cast the slightest doubt in the mind of an injured victim or a grieving survivor or loved one can inflict greater pain upon them than they already are bearing. The security officer's burden is heavy and his or her job may be dirty and dangerous. It may be more than one realized or bargained for, but this should have been taken into account before the person entered the profession. It is too late to be unsure and to feel one made a mistake once an emergency has occurred.

CHAPTER 10 PROBLEMS

PROBLEM 1

The Weeks Atomic Research Complex is a highly secret security plant under control of the Nuclear Regulatory Commission.

All workers require special security clearances. All buildings are totally secured after 5 P.M. and no one is allowed in or out, except in the "C" building, where research personnel often work late into the evening, conducting experiments.

In this building, every door except the front lobby door is secured by special time locks, which can only be released in the event of fire or explosion. These locks are connected to the alarm system.

The lobby door is guarded by an armed security officer who has the only key to the door. The door is locked at 5 P.M. and remains locked until an authorized person seeks entrance. After all identification procedures are completed, the door is unlocked by the guard. The person enters and the door is immediately locked again. The person goes through a sign-in procedure before proceeding further. The door must be unlocked by the guard when a person wishes to leave and after going through a sign-out procedure.

On January 18, Security Officer Sam Wilson is on duty at the lobby post in building "C". At 11:30 P.M., two lab workers, who have been working inside the building, run up to him in an almost panic condition. One of them says, "We heard what sounded like a muffled explosion, or gunshot, on the second floor. It sounded like it came from Dr. Silverman's lab. He's been working there this evening. We didn't go into his lab because it's highly classified and we don't have the right clearance, but we know he was in there. He passed by where we were working about a half hour before and we saw him enter his lab."

Faced with this situation, what do you think Officer Wilson should do? Keep in mind all of the specific details given in the above account.

PROBLEM 2

Officer Mary Moore is walking her post inside the fence surrounding the property of Omega Chemical Company.

The company is on strike. The strike started nine weeks ago and tempers are getting shorter by the day. A striker comes up to the fence, near the spot where Mary is walking, and yells at her.

"Hey lady, how come you do that job for those cheap bums? I'll bet you ain't getting any more than minimum wage. I'll also bet you the men guards get the best shifts and you gals get what's left. How can you be so stupid? That company has been cheating us for years, but we ain't going to take it anymore. Why don't you poor slobs get wise and do what we're doing?"

How should Officer Moore handle this situation?

PROBLEM 3

It is a rainy afternoon and the temperature is about 35 degrees. The wind is blowing and the combination of the rain, wind, and temperature is making it a miserable day for the pickets doing duty at the entrance of Kelley Manufacturing Company. The plant has been on strike for six weeks and there is no end in sight.

Security Officer Frank Smith is in his warm guard house actually feeling sorry for the pickets but keeping his feelings to himself.

The phone in the guard house rings. Officer Smith picks up the phone.

"Main gate guard house, Officer Smith speaking."

"Smith, this is Harold Jones, the plant manager."

"Yes sir, Mr. Jones. What can I do for you?"

"You can get out there and tell those pickets to stay off company property and keep moving or I'm going to call the judge and tell him those people are violating his injunction, and I want them in jail!"

How should Officer Smith handle this?

PROBLEM 4

A large crowd of people, mostly males, is gathered in a huge parking area outside of the entrance to the Brown Chemical Company plant. The crowd is broken up into groups of from 20 to 75 persons each. Most of the people appear to be arguing among themselves, but in one or two groups a single individual has the attention of those in their particular gathering.

Brown Chemical Company is rumored to be making components for chemical warfare projectiles, which the government is shipping to a Middle Eastern nation. This nation is suspected of using the chemicals on the civilian population of its enemy.

The crowd consists of war veterans, college students, and individuals looking for excitement. People started to gather about 8:00 A.M. as a small group. Now the crowd consists of over 1,000 and is still growing.

Security Officer Harold Howard is the day-shift supervisor of a security force of eight men. His officers are stationed at various places through the plant and at the front gate. The front gate is near the parking lot where the crowd has gathered.

Officer Howard is summoned to the gate by one of his guards and made aware of the change in attitude that has taken place among the people in the lot. They seem to have become agitated, irritable, and restless.

What action would you take if you were Officer Howard?

PROBLEM 5

It is Wednesday and Gerry Adams is the security officer assigned to a post at the entrance gate of Artec Corporation. The company has been on strike for two months, and the union pickets walking outside the gate where Adams is stationed are irritable because of the length of the strike and the nasty weather conditions of the last few days.

One of the pickets, seeing Adams opening the gate to let a member of the company management enter the plant, yells in the direction of Adams. "If we ain't got a contract by Friday, we're going to kick down that damn gate and take over the plant. You'd better not be in the way!" he shouts.

What action do you believe Officer Adams should take?

PROBLEM 6

When security officers work a strike for a long period of time there are opportunities to associate and get on friendly terms with personnel of both the union and the management side. This can come about by casually discussing sporting events and other topics of mutual interest not related to the strike, and by accepting coffee and doughnuts, and other favors. Such associations with either side should be:

A. Avoided

B. Encouraged
C. Encouraged only where such association can be established with both sides, otherwise avoided.
D. Neither encouraged nor avoided, but accepted to the extent that these associations develop at the invitation of others than the security personnel.

PROBLEM 7

A large number of demonstrators are marching back and forth in front of several adjacent stores. They are protesting alleged discrimination by the stores. The demonstrators are not blocking any entrances or the passageway of any customers. They are protesting in an orderly manner and are well behaved.

The security officers assigned to the largest of the stores notice the managers of the other stores getting together to discuss the situation. The managers then announce to the crowd that if they don't disperse they will be arrested.

What should the security officers do?

A. Arrest the demonstrators for unlawful assembly.
B. Arrest the demonstrators for blocking traffic.
C. Discuss the grievances with the leader of the demonstrators and the managers, and try to create an atmosphere of understanding.
D. Advise the store managers the demonstration is legal and that interference with it would make them subject to arrest; order nonparticipating individuals who are standing around and gawking to disperse.
E. Furnish the necessary force to permit the managers to make the arrests they have threatened.

PROBLEM 8

The most important objective of security officers when assigned to strike duty is:

A. To protect the rights of workers from unscrupulous employers.
B. To preserve the peace and maintain a neutral position.
C. To protect the employer's property from angry strikers.
D. To control the number of pickets.

PROBLEM 9

What is the most effective way of handling a mob?

A. Use of force.
B. Give in to the mob.
C. Try to reason with the mob.
D. Arrest the leaders.
E. Fire warning shots into the air.

PROBLEM 10

If a crowd, for whatever reason, becomes panic-stricken, the slightest provocation can turn it into an uncontrolled mob. As an uncontrolled mob, it differs from a disorderly or riotous mob, mainly because

A. it is more easily controlled by its self-appointed leaders.
B. often there is more than one leader, and the result is that participants break into controllable groups.
C. each individual acts on his or her own.
D. it is less likely to be motivated by fear, and by the desire to escape.

CHAPTER ELEVEN

Descriptions

A standard method of describing persons and vehicles is used by all law enforcement agencies throughout the United States.

STANDARDIZATION SAVES TIME

The standardization method was developed in order to save time in cases where time is a critical element in a major crime. Most non-law enforcement or nonlegal personnel do not realize or have any conception of how much a single moment, under certain circumstances, can mean. Civilians take time more or less for granted. Seldom in their lives does the loss or gain of a minute or two make a difference to them. That is, of course, unless they happen to be in an athletic contest of some kind. Even then, the worst that can happen is either a win or a loss. They are not usually confronted with the situation where the moment lost or gained can result in a life being saved or lost, especially their own.

Security personnel are seldom faced with this situation in their ordinary line of duty, but the difference between the layperson and the security officer is the fact that it is more apt to happen to a security person. For this reason, if not for other important reasons which exist, it behooves the security officer to pay serious attention to this procedure.

STANDARDIZATION MEANS SAFETY

The standardization method provides an element of safety to the street officer. When the dispatcher communicates information to patrol officers in a roving vehicle, standards in the method of dispatching make detailed and accurate facts available to the officer or officers. It makes available to the officer information that takes some of the edge away from the "bad guys." It provides such descriptive information as is available about the subject or subjects so they become recognizable. They are no longer totally invisible. It removes some of the element of surprise, which favors the criminal. It lets officers know the make, model, color, type, etc., of the vehicle they are looking for, how many suspects are in the vehicle, numbers of males and females, their race and color, and even if they are armed. This information gives the road officer a chance to know what to prepare against and what actions to take. With this information available, detailed plans can be made in advance. Officers know what they are up against and do not have to face a complete unknown factor.

STANDARDIZATION AND REPETITION

By following the same procedure time and time again while giving descriptive information concerning persons or vehicles, a habit is formed. It is a good habit and provides the user with almost immediate recall when required to pass along vital information about persons or vehicles to waiting officers. The information may be required to be transmitted to a number of different agencies taking part in a situation. By using standardized description procedures, each agency knows exactly what the other knows without having to make inquiries. *Time is saved.*

STANDARDIZED DESCRIPTIONS IN REPORT WRITING

The standardized procedure should be used in the writing of reports. By doing so, no matter in what locality or by what agency the report is read, the descriptions will have the same meaning.

By knowing the standards, the mind and eyes focus immediately upon these points when required to do so. It becomes second nature only after a small amount of practice. The almost immediate assistance provided by a complete and accurate description given to the agency involved is invaluable.

DESCRIBING PERSONS

First, consider the standards for describing a person. Although the standards call for specific facts, it is only reasonable to understand that under certain conditions it is not always possible to be exact. Times arise when there must be an estimation made by the person giving the description. Use the system listed below.

1. Name—(if known)
2. Race
3. Sex
4. Age—(If not certain, base an estimate on your own age, or the age of someone your own age, or on the age of someone you know who appears to be the same age. Make the mental comparison and then express your opinion, allowing for a span of from three to five years.)
5. Height—(If not certain, base an estimate on your own height or the height of someone you know who appears to be the same height. Make the comparison and then state your opinion, allowing a span of plus or minus two inches.)
6. Weight—(If not certain, base an estimate on your own weight, or on the weight of someone you know who appears to be the same weight. Make a comparison and then state your opinion, allowing a difference of plus or minus five pounds.)
7. Hair—(Include color, style, length)
8. Eyes—(In addition to color, describe any unusual facts such as unusually large, small, squinty, crossed, etc.)
9. Complexion—(Ruddy, pockmarked, dark skin, light skin, fair skin, blemishes, etc.)
10. Physical features—(Scars, marks, tattoos, one arm, one leg, limp, etc.)
11. Clothing—(Start at the top and proceed downward.)
    a. Hat—(type, style, color)
    b. Shirt, blouse—(style, color, length)
    c. Tie—(style, color)
    d. Coat—(style, color, type, length, material type)
    e. Trousers (or skirt)—(type, color, length, material type)
    f. Socks—(color, length)
    g. Shoes—(type, style, color, material)

DESCRIBING VEHICLES

1. Color—(solid color, color over color; i.e., black over white)
2. Year
3. Make—(manufacturer, model, style name)
4. Body style—(two door, four door, hatchback, pickup, van, convertible)
5. Antenna—(type, number, and location on the vehicle)
6. License plate—(plate number, state of registration: if not available, at least try to furnish plate colors)

There is a simple way to remember this second set of standards. It involves utilizing the first letter of each of the above points of description. You will note they spell CYMBAL. Once you are able to remember this combination of letters and the key word they form, you have no trouble remembering what each letter represents.

DESCRIPTIONS OF OCCUPIED VEHICLES (CRITICAL)

The standards for occupied vehicles is most serious. Be aware of the fact that the following information can mean the difference between life and death for an officer in a patrol vehicle receiving the information from the dispatcher. An officer in pursuit of another vehicle whose occupants are wanted for some crime must know what he or she is up against. Officers need all the information possible in order to plan their approach. Without it, they could be walking into a situation that could result in serious harm or death. *Put yourself in the officer's position—wouldn't you want to know? Wouldn't you like to take away some of the perpetrator's edge?*

1. Wanted for what offense? (critical)
2. Headed in what direction?
3. When last seen.
4. Color of vehicle.
5. Year of vehicle.
6. Make of vehicle.
7. Number of occupants (critical)
8. Sex of occupants (critical)
9. Race of occupants.
10. License plate info
11. Armed; if so, with what and how many? (critical)

CHAPTER 11 PROBLEMS

PROBLEM 1

It is a quiet Sunday evening and in a short time the sun will set.
Security Officer William Harris is at his post outside the cashier's booth in the Heather Shopping Mall parking lot. He's a little shocked at the number of cars in the lot, it being a Sunday.

While thinking about the shopping crowd, his attention is drawn to a gray and black late model Ford bearing down on his post at a high rate of speed. Realizing the occupants don't intend to stop, but to crash the drop gate, Harris ducks behind the cashier's booth for cover. He raises up long enough to see there are three occupants in the four-door vehicle. They appear to be young people. He cannot tell for sure, but the person in the back seat could be a female. The car swerves at a high rate of speed into the westbound lane of Highway 61. The last three license plate numbers may have been 268.

What is Officer Harris's next move?

CHAPTER TWELVE
Report Writing

There are numerous times when security officers are required to write a report concerning their activities. These reports are for employers, employers' clients, and sometimes for the police.

Because most of a security officer's dealings are with a client, an employer, or with the police, it is practical and convenient for the report writing format to be similar or close to uniform to the format used by the police of the jurisdiction in which the security officer is employed. For this reason, this instructional chapter will begin by first examining reports in general. The chapter will then shift to the formats and contents of reports required by police agencies.

The main reason we emphasize the police report style is because of the possibility of an officer's report being needed for use in court during a criminal trial. Keep in mind, the first police officer on the scene is usually responsible for writing the basic or preliminary report. On this basic report may depend the decision as to whether sufficient evidence exists and the presence of the necessary elements of a crime for the issuance of an arrest warrant. This report actually forms the beginning of the state's case for prosecution of a suspect. However, consider the situation in which the first officer on the scene is a security officer. The event takes place within the confines of the client's property where the officer has authority to work. It is only after the incident takes place and is discovered by the security officer that the police are called in. Therefore, the first police officer on the scene only knows what he or she is told by the security officer, or what is in the security officer's report. For this reason, the security officer's report usually becomes a supplement to the first police officer's report, and as such is almost certain to require the security officer to testify concerning the facts in the report. The security officer's report acts as an amendment, just as an investigative report by a detective.

It becomes rather clear, in light of the above explanation, why it is important for security officers to become thoroughly familiar with the style and format of a police report.

The critical examination a police report receives in a courtroom is not the norm for reports to clients and employers, or for other uses.

Just because there may be more scrutiny to the police report than other types of reports, there is no less demand for accuracy, conciseness, neatness, correct spelling and sentence structure, and format in all other reports.

What follows is a general instruction, and although it does provide the information the reader needs to accomplish the task, it will not assure a quality product simply because one reads the chapter. *Quality* in report writing, whether for the courtroom or corporate use, *requires practice.* It's the same as shooting, singing, piano playing, or any other endeavor. If the general principles contained in this chapter are put into practice, applied, reviewed, critiqued, and then reworked, the writer will develop a useful and satisfying skill. You will be able to communicate your thoughts in writing better than most people today. It will provide you with pride in your accomplishment and a strong basis for advancement in your career, whatever it may be.

The main tool necessary for writing a good report is a simple, everyday hand dictionary. This is an absolute must. A good defense attorney loves to have a poorly written report containing a large number of spelling errors. When questioning the first officer on the scene, the lawyer can make incompetence on the part of the officer a factor before the jury. This is a large first step in undermining the state's case against the defendant.

REPORTS IN GENERAL

Reports become records, and records form the foundation of history. Though insignificant at the time of its creation, the report may find its place in the annals of history sometime in the future. It could be involved in some way or another in a case before the U.S. Supreme Court. How sad it would be for such a historical document to be inaccurate or riddled with errors. It would be pitiful to have this happen because its author did not think the report was important, or just didn't care. Little consideration is given this line of thought when an individual begins writing a report. If some thought were given, perhaps the writer would make a more serious effort to do a job that would bring recognition and praise rather than ridicule and scorn.

Reports are written to be accurate, factual presentations of information. It makes no difference what the end use or purpose of the report may be; the basics of construction and content are virtually the same in all reports. Reports, whether used in business, by the courts, by the police, or by individuals, are important. Reports can be the foundation for a system of records, the basis for making important business decisions, or providing the groundwork in making these decisions. Reports can be the basis for legal proceedings, civil or criminal. Regardless of their intended use, reports always contain the central and most important facts concerning an issue at hand or under consideration.

A document of such importance, and having such a place and function in today's social system, is surely worthy of the writer's best effort. Anything less should not even be considered.

It is difficult to estimate the number of uses to which reports are put. They are used to determine actions to be taken. People of authority use them as a basis for promotion, reprimand, or corrective actions. Reports are used to determine progress, to verify facts or conditions, and to carry news. With these and many more uses in mind, it stands to reason that reports can have serious effects and consequences on the lives of a great number of persons. A defective business report, with inaccuracies, errors, or omissions, can result in a civil suit; an incomplete, inaccurate, or otherwise defective police report can lead to a miscarriage of justice.

If every time a person was called upon to write a report he or she would consider the fact that there must be a good reason for the necessity of the report, perhaps that person would take more pride in one's work and provide a worthy result.

REPORT WRITING ABILITY

Not all people have the ability to write a clear, concise, and accurate report. Even with a great deal of practice there are those who never seem to get the knack. Then there are those gifted persons who, with little or no effort, find the task simple and nontaxing. Those who have the ability to compose a well-written, informative report possess an asset invaluable in the courts, in business, in law enforcement, and the entire legal system.

It is a fact that those who have marginal skills in this field can develop with practice, along with trial and error, into effective report writers. When the person reading a report is able to obtain all of the essential facts, no more or no less, in a clear, accurate, concise, and logical order, that person's job is automatically made easier.

Being adept with words and phrases is not enough. Information available to the writer must be analyzed. The important must be separated from the unimportant, or ordinary. The material must be arranged in a manner that not only communicates well but also clarifies and emphasizes those parts that are most important. The resultant report should never be a product of less than the writer's best effort.

Too many persons in business, law enforcement, the courts, and even in simple letter writing give their effort to communicate little concern. If clerks, patrol officers, court clerks, and others wonder why, after 10 years on the job, they are still clerks, patrol officers, or court clerks, they don't have to look far to find their answer.

REPORT CONTENT

Reports vary in content, form, and purpose, but the basic elements and characteristics of all reports are similar. Each has one purpose in common—communicating information and facts from one person to another or one source to another. The information communicated may eventually become part of a record, and there may be a time when that particular record becomes a part of history. One never knows at the moment the report is being written. Because of the importance of what *may be, or even what is, the information contained in the report must be factual, informative, explanatory, critical, analytical, or a combination of any or all of these qualities.*

The material contained in the report is normally the result of observation, research, inquiry, personal knowledge, reason, investigation, and other possible actions on the part of the writer or others. In summary, a report, regardless of the kind or for whom, is an instrument of communication of facts to others containing all of the pertinent information that has been gathered by any number of means.

Since reports seem to require such care and dedication in their preparation, and the necessity for accuracy is so great, the question may be asked, Who needs such a precise document?

Give some thought to where the government, business, educators, law enforcement departments, the judicial system, medical researchers, textbook writers, and others would be without reports. There is hardly a meeting, regardless of its importance, at any level of business in which individuals or committees do not report certain information to the entire body present, to stockholders, to the public, to the press. Based on this commonly known fact, is there really any question why it behooves people in business, government, or other walks of life to learn to write well?

NECESSITIES FOR REPORT WRITING

The following steps make up a critical outline of the necessities of at least adequate report writing.

1. A proper inquiry into the subject material of the intended report.
2. The taking of complete, accurate, transcribable notes.
3. The use of the proper format, depending, naturally, upon whose use the report is intended.
4. Choosing the correct language.
5. Proper use of sentence structure.
6. Complete accuracy of all the facts.
7. No additions or omissions of facts.
8. Distinguishing facts from hearsay, conclusions, judgments, and personal opinions.
9. Clarity.
10. Conciseness.
11. Absolute fairness.
12. Completeness.

BASIC STEPS AND PROCEDURES

There are basic steps and procedures that should be learned in order to write a report which will be structured in such a way as to present a clear and orderly statement of facts.

Report writing is a matter of putting down on paper the pertinent facts concerning an issue or an event, in exactly the same way and in the same order the writer would tell a story were he or she relating it verbally. When telling a story you normally do not tell the ending first. The events are put in chronological order as they occurred. It is really not difficult if some thought is given to the project.

First, make a careful examination of all the information that has been gathered from any and all sources. When reviewing the story in one's mind, the writer should separate the relevant facts and information from that which is not relevant. Then the writer should analyze the relevant material with diligence and with a very critical eye. Having done this, the material must be organized in the sequence that conveys the message intended. At this point it may be helpful, if the writer is not able to visualize an outline in his or her mind, to make one on paper. In this manner the writer can check and double-check the sequences of one's work. It also helps determine whether anything has been omitted. After careful scrutiny, and when satisfied the general organization of facts is to one's liking, it becomes time to compile the final product.

It makes absolutely no difference what purpose the report is intended to serve, or what subject matter is contained therein, the contents must be absolutely factual, unadorned, exact, complete, relevant, and totally accurate. There should be no self-serving additions, or wordy decorations added by the writer. The quality of the final report should be the result of the quality of the inquiry and the investigation. If the inquiry and investigation are poorly done, the resultant report will be the same. The output can be no better than the input. Accuracy, thoroughness, and objectivity during the investigation are key factors.

When finding oneself in a position that requires making a report of events or conditions to others, it would be beneficial to establish a firm foundation upon which to build. This foundation is started by taking notes from the very beginning of the investigation. When taking these notes, the writer should remember nothing is insignificant or unimportant at this stage. The notes must be clear, concise, and accurate. Nothing should be left to guesswork or imagination. Nothing should be so unclear that the writer will not know or understand the meaning of his or her own original notes when it comes to evaluation time.

Make the notes now and evaluate them later. The notes, being the foundation of the report, will make the final product only as good as the notes are. Computer people have a saying, "Garbage in—garbage out." This can also be said of report writing.

THREE PARTS OF A REPORT

If you remember your study of English from high school, you were taught that there are three parts to a letter. The *greeting* or *salutation,* the *body* of the letter, and the *closing.* Report writing is really no different.

Almost all written reports have three main parts. They may assume different names and forms, depending upon the type of report being prepared, and for whom. By and large, the same parts serve the same purpose, regardless of what they are called. The three parts are the *introduction,* the *body* or *narrative,* and the *conclusion.*

There is a major difference between business and other types of reports and a police report. In the police report, the reporting officer never forms a conclusion. It is not the officer's job. That's the job set aside for legal minds and a judge and jury, not the police officer. Officers may write something to the effect that "the driver appeared to be exceeding the speed limit" or "the suspect appeared to be under the influence of some form of medication, drug, or alcohol." Such statements are just a means of describing the appearance of a person's actions. These statements do not state an opinion or a conclusion.

THE FIVE W'S AND AN H

Regardless of the type of report being written, it would be to the writer's advantage to review the final draft by asking the following questions before turning in the report. If there are questions one cannot answer yes to, it would be to the writer's advantage to review what has been written.

First, remember that all reports should answer as many of the following questions as possible. It is not always possible to answer them all because of circumstances, but it is a good place to start the review.

Does the report answer

1. Who?
2. What?
3. Where?
4. When?
5. Why?
6. How?

Can you answer yes to these questions?

1. Have I included all relevant information?
2. Have I omitted all irrelevant information?
3. Have I presented all relevant information in each section and under each heading?
4. Have I presented all details exactly, precisely, and objectively?
5. Have I phrased sentences clearly, precisely, and objectively?
6. Have I used proper paragraph form?
7. Have I used proper mechanics?
8. Have I communicated to the reviewer an accurate, complete, and clear understanding of the case, as a whole and in its parts?

THE POLICE REPORT AND ITS SECURITY COMPANION

Since security work is so closely related to police work, and since there are times when the users of the reports will be the same audience, it seems only proper that in this book we should stress the police report because of its form and how by using it as a guide we can wind up with a solid security report.

Police reports are used for criminal trials.
Police reports are used for interdepartmental police messages.
Police reports are often used in civil trials.
Police reports are used by insurance companies.
Police reports are used by background investigators.
Police reports are used by the media.
Police reports are used by prosecution and defense lawyers.

The majority of uses outlined above are possibly shared by the security officer's report. In addition, the security report may be used in reports to clients regarding,

1. Violations of company rules and regulations.
2. Accidents or injuries on the property.
3. Poor safety conditions and violations of safety rules.
4. Accounts of trespassers.
5. Possible fire hazards.
6. Poor housekeeping.

IMPORTANCE OF THE REPORT

The written report is a communication document. It sends a message from the officer writing the report to appropriate officials who will read the document and take further action as a result of the contents of the report. Decisions to prosecute a crime will be made only if the report contains adequate, accurate information. In the case of a police report, and some security reports, these documents become a permanent, official record of the police department, and in many cases of the criminal or civil court system.

It is critical for an officer to be aware of the fact that criminal prosecution actually has its genesis in the original police report, and is normally based on its contents. The contents of the report determine whether there is sufficient evidence and whether all of the necessary elements of a crime are present to justify the issuance of an arrest warrant. Later it will be the foundation upon which the prosecution will build its case against the person arrested.

The prosecution must be given clear, concise, truthful, and accurate facts. It must have a well-organized, grammatically correct report, free of punctuation and spelling errors. To go into a court of law with any lesser document is to invite disaster. Any good defense attorney will use a sloppy, ill-prepared report to tear down the competency of the officer who wrote it. In the eyes of the jury, the officer, usually the first person on the scene, is a major witness and is critical to the trial. To destroy his or her credibility or competency is to rock the foundation of the prosecution's case.

Explaining the Three Major Parts of a Police or Security Report

Earlier we said the three parts of a report are the introduction, the body, and the conclusion. To illustrate, consider the police report. The purpose of the *introduction* is to provide a clear definition of the subject matter, the scope of the document, and pertinent information concerning the major actors involved. If a department or company uses a cover sheet or face sheet, in its reports, this normally fulfills the same function. The introduction contains the name or the type of crime being reported, the names and descriptive information concerning the victim or victims, the witness or witnesses, and any or all suspects. This information permits the reader immediate access to the personal details and descriptions of all of the principals contained in the report. It also gives readers the opportunity to familiarize themselves with these persons before entering into the body of the report. The reader is able to form an image of those known to be involved so he or she is better able to visualize the events taking place in the body content by first knowing something about the cast of characters.

The *body* of the report contains the facts and findings that have been obtained as a result of investigation or the witnessing of the events by the person making the report, or by any other witnesses. It may contain information in many forms, such as photographs, sketches, charts, drawings, quotations, or verbal descriptions, all of which tend to enhance the other contents. The meat of the report is found in the *body.* This is the reason so much care must go into its preparation and construction.

The *conclusion* of the document follows immediately after the body. It may contain a summary of the preceding events.

Report Contents and Questions Needing Answers

Reports, other than those used by the police and law enforcement agencies, may require the recommendations, conclusions, suggestions, or ideas of the writer based on the facts contained in the report. This can be an important part of the private industry report.

The police or security report should contain nothing but facts. They should be written in chronological order as the event or events unfolded. The report should be clear and concise, free of superfluous rhetoric. Care should be taken so that important details are not omitted.

The method of preparing a police or security report is simple. Review the incident in your mind exactly as it took place. Make notes, answering questions referred to as the Five W's and an H. The answers to these questions can provide the outline from which the writer is able to enlarge upon. The questions fall under the general categories of *who, what, when, where, why,* and *how.* If the answers to these questions can be found in the report, provided they are known to the writer, of course, he or she has written a successful report.

A close look at the same checklist of questions under each of these categories will assist the writer as to the type of material that is important and necessary. The answers should permit the writer to form thoughts about the contents of the body of the report. From that the writer can determine where in the sequence of things they belong.

Many more questions can be asked under each category, but the sample list that is provided below will give you an idea of some questions and how the answers to them can help you write a complete and thorough report.

Start the checklist by answering questions in the *Who* category. Add your own list of questions if you think of anything you may need answered under the category. If the questions can be answered, write these answers in a note book or on a note pad. Hold them until all of the questions in all of the categories are answered, and then make your outline.

Who is the victim of the crime?
Who committed the crime?
Who witnessed any part of the crime or activity?
Who were accessories?
Who stood to profit from the crime?
Who furnished information or any tips?
Who are suspects?
Who discovered the crime?
Who was the first officer on the scene?
Who made the initial investigation?
Who assisted?
Who was interviewed or questioned?
Who protected the crime scene?
Who processed the crime scene, gathered or preserved the evidence?
Who did the interviewing or interrogating?
Who made identifications of whom or what?
Who made arrests?
Who canvassed the area for leads?
Who made any phone calls?
Who transported victims, suspects, or evidence?
Who drove which vehicles?
Who destroyed evidence?

It should be evident that a list of such questions can extract many answers concerning material necessary for incorporation in a detailed, complete report. Add to this list many of the *who* questions that fit the particular crime or event and the "report body content" will begin to expand.

Follow the *who* list with questions concerning *what.*

What is the type of crime?
What took place?
What are the injuries or damages?
What was taken?
What is the estimated value?
What statements were made by the victim, the witnesses, the suspect?
What actions were taken by anyone?
What evidence was obtained?
What types of vehicles were involved?
What is the description of the vehicle, the suspect, the witnesses?
What was the time?
What was the cause?
What were the excuses?
What was the reason?
What type weapon was used?

This list of *what* questions can and will be added to as the writer becomes familiar with the process and how the answers fit into his or her mental image as the writer reviews the entire incident in his or her mind.

Start by asking the next question—*when?*

When was the crime committed?
When was the crime discovered?
When was the crime reported?
When did the first officer arrive?
When did the ambulance arrive?
When did others arrive?
When was the subject apprehended?
When was the coroner notified?
When was the suspect first seen?
When did the witness arrive or leave the scene?
When was the body removed?
When was the last inventory taken?
When were the shots fired?

It is always important for major movements and actions to be recorded within a time period, even if approximations are necessary.

The answers to *where* will fix locations that can be extremely important in certain cases.

Where did the crime take place?
Where was the victim?
Where was the suspect?
Where do various persons live?
Where were various persons last seen?
Where were various persons when the crime took place?
Where was the material taken from?
Where was the evidence discovered?
Where were the interviews and interrogations held?
Where were the various persons going?
Where did the suspect obtain the weapon?

Answers to some *how* questions are readily available, and to others the answers depend upon the result of the investigation.

How was the crime committed?
How was it reported?
How was the victim killed or injured?
How did the victims, suspect, or witness get to the scene?
How was anyone moved?
How was anyone dressed?
How were the valuables removed from the scene?
How was the entry made?
How much was taken?
How often did something happen?
How often was the victim shot, stabbed, hit?
How did the suspect escape?

Why questions are usually a little more difficult to answer and sometimes the answers can only be found as a result of a long investigation.

Why was the victim killed?
Why was the suspect in the area?
Why did the dog bark?
Why was it the only object of value taken?
Why didn't they take the easy way out?
Why was the victim in the area where he or she was killed?
Why didn't the victim's husband or wife tell the truth?
Why did the suspect run?

After considering all of the answers obtained from these and similar questions, it can be readily seen that by putting these answers into an outline according to sequence of importance, a complete picture can be envisioned and a thorough report made, communicating a vivid story to the reader.

REVIEWING EXHIBITS

With the previous instruction on report construction in mind, review the sample police report marked "Exhibit A" (see pp. 99–102).

Sheets 1 and 2 of Exhibit A provide the introduction or first part of the report. The sheets make up the first two pages of the "uniform crime report" form. You will note that page 1 introduces the reader to the type of incident involved in the report, namely "Armed robbery." By providing this information on the introduction sheet, the reader is not required to read through the entire report to determine what crime took place. It also provides the means and the opportunity to classify the report. Then, in summary form, sheets 1 and 2 give the cast of characters involved in the incident, along with pertinent descriptive and informative data concerning each person. The amount and type of physical evidence is also listed so the information is readily available to investigative officers, the prosecuting attorney's office, and the courts.

Sheets 3 and 4 of Exhibit A contain the *narrative* or *body* of the report. The body relates the facts concerning the incident, in simple yet clear and complete detail. The story is related step by step as the incident took place, by the reporting officer. While reading the body of this report, it is almost possible for readers to put themselves in the place of the reporting officer as the events unfold. There is nothing glossy added to make anyone sound heroic or to attract individual attention.

The "conclusions" on page 4 of Exhibit A consist of a simple statement advising "further information will be added by supplemental reports." Because this is a police report, this section of the document contains no opinions, no individual thoughts, or conclusions, and no recommendations. These efforts are usually only included in business or medical reports.

When referring to Exhibit B (which follows Exhibit A), however, we are dealing with a different matter. Security personnel do not normally have such a detailed report form at their disposal, unless their company has gone to the effort of designing one. (Exhibit C is the design of a cover sheet for a security officer's report. It encompasses the information needed and yet is not as all-inclusive as the police form.)

For this reason the task of incorporating the necessary information and answers to questions falls directly on the shoulders of the individual security person writing the report.

Exhibit B gives an idea of the type of report a security officer may submit. It tells the story of an incident that took place during the course of a guard's normal work shift. The story is told in clear terms and chronological order, and it answers the requirements of the Five W's and an H where and whenever possible.

Special note should be taken of the manner in which the various participants in the incident have been "pedigreed" or described. In the case of John Smith and Mary Adams, they are considered to be in the category of witnesses. For this reason, a little more information is necessary concerning the specific details of their lives than that of the victim. Witnesses may have to be contacted at some future date, and they are not usually as available as the victim. Having the person's occupation, name, and address of their workplace, along with their business phone number, makes getting in touch with them much easier should they not be at home when contacted.

Finally, the only way to become proficient in report writing is *practice, practice,* and more *practice.* But as practice *writing* is taking place, practice *spelling* should also take place. A report filled with spelling errors is the easiest way in the world for a security officer to lose stature in the eyes of a client, an employer, an attorney, and/or the court.

STAPLE HERE

POLICE DEPARTMENT ANYWHERE U.S.A.

LAW ENFORCEMENT OFFENSE / INCIDENT REPORT

1 COPIES TO	2 OFFICER INJURED	3 PROCESSED BY	4 CODE	5 REPORT FOR	6 PAGE	7 COMPLAINT NUMBER
File	☐ YES ☒ NO		21		1 OF 4	87-63281

8 TYPE OF INCIDENT	9 SICC	10 UCR CODE	11 HOW COMPLAINT WAS RECEIVED (Circle)	12 ARRIVAL TIME
ARMED ROBBERY			X 1 RADIO 2 CITIZEN 3 PHONE 4 STATION 5 ON VIEW	10:32 p.n.
	13 ORIGINALLY RECEIVED AS: Armed Robbery in Progress		14 OTHER UNITS NOTIFIED (DSN AND UNIT NAME): Unit 328 Smith,J DSN1110,Williams,JDSN989	

15 DAY, DATE, TIME OF OCCURRENCE	16 DAY, DATE, AND TIME REPORTED	17 STATUS
Novembef 8,1987 10:25 p.m.	11-8-87 10:27 p.m.	☐ ACTIVE ☐ INACTIVE ☒ CLEARED BY ARREST ☐ EXCEPTIONALLY CLEARED ☐ UNF

18 LOCATION OF OCCURRENCE SPECIFY STREET ADDRESS AND/OR APARTMENT COMPLEX, SUBDIVISION, HOTEL	19 CAR/BEAT	20 DISTRICT	21 GEOGRAPHIC CODE
1065 Adams St.	326	2	2

22 TYPE OF PREMISE (Drugstore, Garage, etc.)	23 POINT OF ENTRY	24 POE VISIBLE TO PATROL
Liquor Store	Front Door	YES ☒ NO ☐

25 METHOD OF ENTRY	26 TOOLS USED	27 POINT OF EXIT	28 WEAPONS/OBJECTS USED OR DISPLAYED
Walked In	none	Front Door	38Spec Blue St/Revolver

29 VICTIM (If Firm, Name & Type of Business) CHECK MULTIPLE VICTIMS ☐	30 RESIDENCE ADDRESS	31 RESIDENCE PHONE	AREA CODE
Leland, Bert	8120 Arch St.	321-8555	321
	32 BUSINESS ADDRESS: 1065 Adams St.	33 BUSINESS PHONE: 414-6200	321

VICTIM'S PEDIGREE	34 RACE	35 SEX	36 AGE	37 DATE OF BIRTH	38 PLACE OF BIRTH	39 OCCUPATION	40 MARITAL STATUS
	W	M	24	5-9-63	Arkansas	Clerk	M

41 VICTIM CONVEYED TO	42 VICTIM CONVEYED BY	43 HOSPITAL DISPOSITION	44 TYPE OF INJURIES
N/A	N/A	☐ ADMITTED ☐ LEFT FOR TREATMENT ☐ TREATED AND RELEASED ☐ DOA	☐ FATAL ☐ DISABLING ☐ EVIDENT, NOT DISABLING ☐ PROBABLE, NOT APPARENT ☐ NOT APPARENT ☐ UNKNOWN

CODES: 1 - Reporting Party 2 - Person Securing Premise 3 - Witness 4 - Owner 5 - Title Holder 6 - Person Discovering Crime 7 - Last Person in Possession 8 - Guardian 9 - Parent 10 - Property Insurance

45 CODE	46 NAME (Last Name First)	47 ADDRESS	48 PHONE NO. AREA CODE
45A	N/A	47A RESIDENCE / BUSINESS	48A
45B	N/A	47B RESIDENCE / BUSINESS	48B
45C 10	INSURANCE AGENT / INSURANCE COMPANY	47C BUSINESS / BUSINESS	48C

VEHICLE CODES: N/A 1 - Stolen 2 - Used 3 - Wanted 4 - Held as Evidence 5 - Victim's Vehicle 6 - Recovered 7 - Bicycle 8 - Towed #

49 CODE	50 YEAR	51 MAKE	52 MODEL	53 STYLE	54 TYPE	55 COLOR	56 I.D. NUMBER
N/A							

57 LICENSE NUMBER	58 STATE	59 YEAR	60 PLATES MISSING	61 KEYS IN VEHICLE	62 IGNITION LOCKED	63 DOORS LOCKED	64 INSURED	65 VALUE	66 REFERENCE NO
			☐ YES ☐ NO No.	☐ YES ☐ NO ☐ UNKNOWN	☐ YES ☐ NO ☐ UNKNOWN	☐ YES ☐ NO ☐ UNKNOWN	☐ YES ☐ NO ☐ UNKNOWN		

PROPERTY CODES: A—Currency Notes B—Jewelry/Precious Metals C—Clothing, Furs D—Damaged Property E—Office Equipment F—TV, Radio, Cameras, etc. G—Fire Arms H—Household Goods I—Consumable Goods J—Live Stock K—Misc. L—Tools

67 TOWED TO ADDRESS:

68 CODE	69 QUANTITY	70 PROPERTY DESCRIPTION: Brand Name, Serial Number, Model/Style, Oper. I.D./Other Identification	71 VALUE	72 RECOVERED	73 REFERENCE NO.
A	1	Ser.G28123B54C U.S.Currency	$50.00	Yes	
A	1	Ser.H234134436 U.S.Currency	20.00	Yes	
A	L	Ser.G063510664 U.S.Currency	10.00	Yes	
A	1	Ser.E838079743 U.S.Currency	5.00	Yes	

74 NARRATIVE	75 TOTAL	76 TOTAL	77 OPERATION IDENT.
See Narrative Sheet	$85.00	$85.00	☐ YES ☐ NO

83 COMPLAINT NUMBER

78 SUBMITTED BY OFFICER, DSN	79 DATE OF REPORT	80 MESSAGE NO	81 SUPERVISOR'S INITIAL	82 REVIEWING OFFICER'S SIGNATURE AND DSN
Ptm.Patterson, James DSN 1000	11/8/87	-		

FIM

Exhibit A

STAPLE HERE

POLICE DEPARTMENT ANYWHERE U.S.A.

PERSONAL DESCRIPTORS CONTINUATION FORM

1 PAGE	2 COMPLAINT NUMBER
2 OF 4	87-63281

CODE — PREFIX: A - ARRESTED; R - RUNAWAY; W - WANTED; J - JUVENILE; M - MISSING PERSON. SUFFIX: 1 - ARMED & DANGEROUS; 2 - ASSAULTS OFFICERS; 3 - MENTAL CONDITION

SUBJECT NO. —

3 CODE	4 LAST NAME / FIRST NAME / MIDDLE NAME / JR./SR.	5 ARREST REG. # /JUV. CONFID. HISTORY #
A	Bekins,William J.	87-126

6 AKA OR MAIDEN NAME	7 MIRANDA WARNING	8 REFERENCE NUMBER
"Bones"	[X] YES [ ] NO DSN	N/A

9 RACE	10 SEX	11 AGE	12 DATE OF BIRTH	13 PLACE OF BIRTH - CITY / STATE	14 MARITAL STATUS	15 CLOTHING
W	M	30	6-10-47	Anywhere U.S.A	S	Blue Levi Jacket Blue Levi Jeans

16 HEIGHT	17 WEIGHT	18 BUILD	19 COMPLEX.	20 EYES	21 HAIR	22 FACIAL HAIR	23 TEETH	24 SCARS, MARKS, TATOOS
6'	190	stocky	Ruddy	Gry.	Bwn.	None	Good	None

25 DEFORMITIES	26 VOICE	27 SOCIAL SECURITY NUMBER	28 OPERATOR'S LICENSE NUMBER	29 STATE
None	Normal	000— 11 —000	None	None

30 ADDRESS	31 CITY	32 STATE	33 HOME PHONE (AREA CODE)	34
655 Apple Ave.	Anywhere	U.S.A	None	[ ] EMPLOYED [X] UNEMPLOYED

35 OCCUPATION	36 EMPLOYER (PRESENT OR LAST)	37 BUSINESS ADDRESS	38 CITY	39 STATE	40 (A.C.) BUSINESS PHONE
Unemployed	N/A	N/A	N/A	N/A	N/A

41 CHARGE(S)	F M / C D	42 DATE OF OFFENSE	43 CN B/W UTT CCT. SUMS.	44 COURT	45 COURT DATE	46 TIME	47 COMPUTER ASS'T ARREST
A Armed Robbery	F M C D	11-8-87	N/A	N/A	N/A	N/A	[ ] YES [X] NO
B	F M C D						[ ] YES [ ] NO
C	F M C D						[ ] YES [ ] NO
D	F M C D						[ ] YES [ ] NO

SUBJECT NO. —

48 CODE	49 LAST NAME / FIRST NAME / MIDDLE NAME / JR./SR.	50 ARREST REG. # /JUV. CONFID. HISTORY #

51 AKA OR MAIDEN NAME	52 MIRANDA WARNING	53 REFERENCE NUMBER
	[ ] YES [ ] NO DSN	

54 RACE	55 SEX	56 AGE	57 DATE OF BIRTH	58 PLACE OF BIRTH - CITY / STATE	59 MARITAL STATUS	60 CLOTHING

61 HEIGHT	62 WEIGHT	63 BUILD	64 COMPLEX.	65 EYES	66 HAIR	67 FACIAL HAIR	68 TEETH	69 SCARS, MARKS, TATOOS

70 DEFORMITIES	71 VOICE	72 SOCIAL SECURITY NUMBER	73 OPERATOR'S LICENSE NUMBER	74 STATE
		— —		

75 ADDRESS	76 CITY	77 STATE	78 HOME PHONE (AREA CODE)	79
				[ ] EMPLOYED [ ] UNEMPLOYED

80 OCCUPATION	81 EMPLOYER (PRESENT OR LAST)	82 BUSINESS ADDRESS	83 CITY	84 STATE	85 (A.C.) BUSINESS PHONE

86 CHARGE(S)	F M / C D	87 DATE OF OFFENSE	88 CN B/W UTT CCT. SUMS	89 COURT	90 COURT DATE	91 TIME	92 COMPUTER ASS'T ARREST
A	F M C D						[ ] YES [ ] NO
B	F M C D						[ ] YES [ ] NO
C	F M C D						[ ] YES [ ] NO
D	F M C D						[ ] YES [ ] NO

MISSING PERSON / JUVENILE

93 HAT	94 COAT	95 SHIRT/BLOUSE	96 TROUSERS	97 SHOES	98 DRESS	99 SKIRT

100 DATE LAST SEEN	101 TIME LAST SEEN	102 LOCATION LAST SEEN

103 JEWELRY, PAPERS OR OTHER ITEMS CARRIED

104 POSSIBLE CAUSE OF ABSENCE	105 PROBABLE DESTINATION	106 HAS SUBJECT BEEN REPORTED MISSING BEFORE? DATE:
		[ ] YES [ ] NO

107 WHERE WAS SUBJECT LOCATED AT THAT TIME?	108 ACCOMPANIED BY:

109 RELIGION - CHURCH OR FRIENDS	110 SCHOOL	111 WILL PARENTS CALL FOR?
		[ ] YES [ ] NO

112 SUBMITTED BY OFFICER, DSN	113 DATE OF REPORT	114 MESSAGE NO	115 SUPERVISOR'S INITIAL	116 REVIEWING OFFICER'S SIGNATURE AND DSN
Ptm.Patterson, James DSN1000	11/8/87			

Exhibit A (continued)

IF ADDITIONAL SPACE IS NEEDED USE ANOTHER CONTINUATION FORM CC

1. DEPARTMENT REPORTING	FORM NO F-9L	2. DEPARTMENT FILE NO
Anywhere U.S.A.	CONTINUATION	87-63281
3. DATE OF THIS REPORT: 11-8-87	DETAILS STOLEN PROPERTY PERSONS WANTED - ARRESTED - VICTIM - WITNESS	PAGE 3 OF 4 PAGES
5. VICTIM OR COMPLAINANT: Leland, Bert	6. PLACE OF OCCURRENCE: 1065 Adams St.	

Sir:

While on patrol in unit 326, I was dispatched by radio at 10:30 p.m., this date to the Adams Liquor Store, 1065 Adams Street, to answer a call of an Armed Robbery in Progress. (Turned in by silent alarm). I proceeded, code 2, to the assigned address, arriving at 10:32 p.m.

I parked my unit on Adams Street, at the curb, approximately 50 yards north of the Adams Liquor Store. I approached the front of the store, with revolver drawn, staying out of view of the store front window.

As I was approaching the building, unit 328, a two man unit with Officers James Smith, DSN 1110, and William Henry, DSN 1121, arrived on the scene as back up. Officer Smith parked his unit at the street curb, approximately 50 yards south of the liquor store. He approached , carrying a **riot gun**. Officer Henry went to the rear of the building with his revolver drawn, to cut off any rear exit escape.

I looked into the store from the corner of the front window and observed the store clerk, Bert Leland, standing behind the counter at the rear of the store. He was next to the cash register with his hands in a raised position over his head. The suspect was standing in front of the counter, at the cash register. The suspect was identified as:

Bekins, William
w/m
6'(approx)
190 lbs.(approx)
brown hair, shoulder length
ruddy complextion
Levi jacket, blue
Levi pants, blue (faded)

Suspect Bekins was holding a blue steel revolver in his right hand, pointed at the victim. He reached around the register and into the cash drawer with his left hand. The suspect ordered the victim to lay on the floor after taking the money from the register. When the victim complied, the suspect headed for the front door. Officer Smith and I

Exhibit A (continued)

IF ADDITIONAL SPACE IS NEEDED USE ANOTHER CONTINUATION FORM CC

1. DEPARTMENT REPORTING	FORM NO F-9L	2. DEPARTMENT FILE NO
Anywhere U.S.A.	CONTINUATION	87-63281
3. DATE OF THIS REPORT	DETAILS STOLEN PROPERTY PERSONS WANTED - ARRESTED - VICTIM - WITNESS	4.
11-8-88		PAGE 4 OF 4 PAGES

5. VICTIM OR COMPLAINANT	6. PLACE OF OCCURRENCE
Leland, Bert	1065 Adams

waited until the suspect exited the front door and was on the sidewalk in a position between us. At this time I called out in a loud voice, "Police, drop the gun." The suspect came to a stop, hesitated as if trying to locate my position. Officer Smith then called out, "Police, drop the weapon." Hearing a second voice on his other side, the suspect dropped his gun and raised his hands over his head.

Officer Smith and I approached the suspect. I handcuffed the suspect's wrists behind his back and gave him his Miranda warnings. Officer Smith recovered a blue steel Smith and Wesson 38 special 4" barrel, Model 10 revolver, ser #C-832608 and eighty five dollars ($85.00) U.S.Currency from the left pocket of the suspect's jacket. (1-$50.00 bill, 1-$10.00 bill, 1-$20.00 bill, and 1-$5.00 bill).

I called the dispatcher by radio and requested an identification unit at the scene. Identification unit #330, with officers Henry Clay, DSN1181 and William Chapman,DSN999 arrived on the scene at 10:55 p.m., and began processing the scene (store area).

The revolver and the money were turned over to Officer Clay to be marked and identified as evidence. I asked victim Leland to come to the station after closing the store to take his statement. Officers Smith and Henry transported the suspect in unit 328 to the police station where he was booked by Sgt.Robert Jones, DSN788. He was booked for "Armed Robbery" and held pending application of warrants in the morning.

At 11:45 p.m., victim Leland arrived at the station and gave his statemtent.

Further information will be added by supplemental reports.

Respectfully submitted,

Ptmn.Patterson, James DSN 1000

Exhibit A

REPORT SIR: SHEET 1 OF 2

WHILE MAKING MY ROUTINE ROUND OF THE PARKING LOT
OF ST. BENEDICTS HOSPITAL AT APPROXIMATELY 8:30 P.M. THIS
DATE, I WAS APPROACHED BY

SMITH, JOHN W/M
D.O.B. 5-9-60
AGE 27
ADDRESS 1234 LONE STAR AVE.
PHONE 356-0010
OCCUPATION CARPENTER
BUSINESS ACE CONSTRUCTION CO.
BUS. PHONE 872-6666

WHO STATED HE WAS GOING TO HIS CAR AFTER VISITING
A FRIEND IN THE HOSPITAL WHEN HE NOTICED A WHITE
FEMALE IN A SITTING POSITION AT THE NORTH END OF THE
PARKING LOT, ACTING PECULIAR, AS IF SHE HAD BEEN
STUNNED. I, ALONG WITH MR. SMITH, PROCEEDED TO THE
AREA DESCRIBED AND THERE FOUND

JONES, WILLMA W/F
DOB. 10-10-20
AGE 67
ADDRESS 821 SOUTH 8TH ST.
PHONE 862-1212

SITTING ON THE GROUND IN A DAZED CONDITION. I REQUESTED
MR. SMITH TO REMAIN WITH THE VICTIM WHILE I SOUGHT
MEDICAL HELP FROM THE HOSPITAL EMERGENCY STAFF.

ADAMS, MARY - NURSE W/F
D.O.B. 6-1-60
AGE 27
ADDRESS 4631 BOLAND DR.
PHONE 451-8873
OCCUPATION NURSE
BUSINESS ST. BENEDICTS HOSPITAL.
BUS. PHONE 311-0100

RESPONDED TO THE SCENE. SHE ATTENDED TO THE VICTIM AND
FOUND HER TO BE IN A MILD STATE OF SHOCK WITH
NO PHYSICAL INJURIES. NURSE ADAMS, USING A WHEEL CHAIR
TOOK THE VICTIM TO THE EMERGENCY ROOM TO LET HER

REPORTING OFFICIAL	RANK	SHIFT	SIGNATURE
WILLIAM CALBOT	PTMN.	2ND	William Calbot

DATE	INDORSING OFFICIAL
11/1/87	Henry Williams

Exhibit B

CONTINUATION SHEET

SHEET 2 OF 2

RECOVER HER COMPOSURE. AFTER ABOUT 1 HOUR, I WAS ABLE TO SPEAK TO THE VICTIM. SHE ADVISED SHE WAS WALKING TO HER CAR WHEN A

WHITE OVER RED
1970
DODGE
TWO DOOR
LICENSE UNKNOWN

SPED PAST AND ALMOST STRUCK HER. SHE JUMPED OUT OF THE WAY AT THE LAST SECOND BUT SUFFERED A SERIOUS SCARE. I NOTIFIED THE LOCAL POLICE BY PHONE, SPEAKING TO SGT. MEL ANDERSON. I THEN NOTIFIED THE VICTIM'S HUSBAND WHO CAME AND GOT HIS WIFE.

REPORTING OFFICIAL	RANK	SHIFT	SIGNATURE

DATE	INDORSING OFFICIAL

Exhibit B (continued)

INCIDENT REPORT

Copies to:	Off. Injured? Y N	Report No.	Type of Incident	Report Date	Page of

Time of Incident	Arrival Time	How Received Radio Citizen Phone On View	Date of Incident	Date and Time Reported	Location of Incident Street Address or Area

Type of Premise	Point of Entry	Method of Entry	Point of Exit	Weapons Used Y N	Type

Victim	Home Address	Home Phone
	Business Address	Business Phone

Race	Sex	Age	DOB	Place of Birth	Marital Status	Occupation

Type of Injuries	Conveyed To	Conveyed By

Code #	1-Reporting Party	#2-Person Securing Prem.	#3-Witness	#4-Owner	#5-Discoverer

	Name	Home Address	Home Phone
		Business Address	Business Phone
	Name	Home Address	Home Phone
		Business Address	Business Ph one
	Name	Home Address	Home Phone
		Business Address	Business Phone

Vehicle Codes #1-Stolen #2-Used #3-Wanted #4-Held as Evidence $5-Victim's Vehicle $6-Towed

Code	Year	Make	Color	Style	Type	Color	Vin No.	Plate No.	State

Suspect Name	Aka	Race	Sex	Age	DOB	POB	DL#
	Mar.Status	Hgt.	Wgt.	Build	Complexion	Eyes	SS#
	Hair	Hair Style	Facial Hair	Teeth	Scars and Tattoos		

Suspect Name	Aka	Race	Sex	Age	DOB	POB	DL#
	Mar.Status	Hgt	Wgt.	Build	Complextion	Eyes	SS#
	Hair	Hair Style	Facial Hair	Teeth	Scars and Tattoos		

Reporting Officer	Shift	Date	Client

Exhibit C

CHAPTER 12 PROBLEMS

PROBLEM 1

The stores inside the shopping mall were just closing, and Security Officer Henry Watson was making his last round of the evening. He looked at his watch and saw it was 14 minutes past 10. It was Saturday night and he was thinking of going fishing the next morning with his son.

His thoughts were interrupted by what sounded like two quick gunshots. Looking in the direction of Macon's Drug Store, from where the shots seemed to come, he saw two men run out of the store. Watson was about 25 feet from the store. The parking lot lights were still on and it made visibility quite good.

Officer Watson was about to draw his pistol when the one fleeing suspect, a white man about 6 feet tall and weighing about 180 pounds, turned and saw Watson. He raised what appeared to be a nickel-plated revolver and fired a shot. Watson dropped to the ground when he saw the man point the gun at him. The shot missed and Watson drew his own revolver. As he was about to shoot, the store cashier came running out of the store screaming, "We've been robbed!" Fearing that he might hit the woman, Watson held his fire.

During the instant the robber was looking at Watson and preparing to fire, the security officer noticed that the man had a dark beard and mustache, a thin face, and was about 30 to 35 years of age. He was wearing blue jeans, a red and green plaid shirt, and a blue stocking cap.

Watson couldn't see the face of the second man, but determined he was younger than the first. He believed the man to be white, between 20 to 25 years of age, stocky build, weighing about 200 pounds, and about 5'10" tall. He was wearing a red baseball cap, white T-shirt, and blue jeans. The second man did not appear to be armed.

Both men ran to a waiting car, a 1985, 4-door, black Dodge Aries K. The car was too far away to see the license plate. The vehicle left the parking lot at a high rate of speed and turned north into 11th Street.

Watson ran into the drug store, saw no one was injured, and called the police. He gave them all of the pertinent information in order to get the message to any police cars in the area.

Watson then went to the cashier, asked her name, and other information. She was Miss Nancy Weber, 8120 Selma Ave., Austin, Missouri, white female, DOB 10/10/76, home phone 330-2126.

Write a report of this incident as if you were Officer Watson.

PROBLEM 2

When a security officer has not completed a special report within the time specified by his supervisor he should

A. without consulting his supervisor, work overtime to complete the report.
B. turn in whatever is completed.
C. turn in all the work in its incomplete form.
D. advise his supervisor of his progress and receive further instructions from the supervisor.

PROBLEM 3

Security Officer Jill Howard has written a report of an incident that took place near her post and is about to turn it in. She discovers some additional information she hadn't known at first. Whether or not she rewrites her report to include this additional information should depend mainly upon the

A. amount of time left in which to submit the report.
B. effect this information will have on the conclusions of the report.
C. number of changes she will have to make in her original report.
D. possibility of turning in a supplementary report later.

PROBLEM 4

For effective presentation, the written report must be adapted to the

A. reader's point of view.

B. reader's experience and knowledge.
C. reader's method of thinking.
D. nature of the subject matter.
E. all of the above.

PROBLEM 5

In the preliminary preparation of a written report, the writer should consider

A. the purpose.
B. the reader.
C. the circumstances and limitations.
D. use of the report.
E. all of the above.

PROBLEM 6

In writing a report it is not important to remember that

A. understanding one's own report guarantees the report will be understood by others.
B. a report of an investigation is usually better if the person preparing it thinks over the subject for a week or two before writing the report.
C. a brief outline of the points to be covered should be made before writing or dictating a report.
D. it is necessary to avoid repeating information.

PROBLEM 7

One way to make written material more understandable to the reader is to avoid "big" words. Another good way is to

A. repeat all important words.
B. include only one idea in each paragraph.
C. use short sentences.
D. avoid using pronouns.
E. clearly define all terms.

PROBLEM 8

The chief reason why it is good procedure to prepare an outline of the points to be covered at the first step in writing a long report is that

A. this is an excellent means of creating a favorable impression with one's supervisor.
B. the writer is thus forced to write the final report in fewer words.
C. such procedure is more business-like and efficient.
D. necessary additions or changes can be made more easily and quickly on an outline than in written material.
E. an outline makes it easy to see exactly how far along the report is at any given time.

PROBLEM 9

The most important requirement in report writing is

A. neatness.
B. grammatical construction.
C. accuracy.
D. promptness in turning out reports.

CHAPTER THIRTEEN

Crime Scene Protection

Too many cooks can ruin the soup. Chances are you have heard this saying many times. The statement is especially true in the case of the scene of a crime.

There is another saying within the law enforcement family: "A criminal always leaves something at the scene of his crime." Whatever it is the criminal leaves may be difficult to find. It may not be in view or visible to the naked eye, but something has been left behind. This being the case, the police evidence technicians and detectives must determine what it is they are looking for. They must then tie the pieces of evidence to the alleged perpetrator of the crime.

Detectives and investigators work backward in time. The more time that elapses between the commission of the crime and the apprehension of the guilty party, the more difficult it is to prove the case. Clues must be developed and worked on as soon as they become available. Time is all-important. There can never be too many clues or pieces of evidence for the prosecution's case against an alleged criminal.

TYPES OF EVIDENCE

Physical evidence, that is, evidence consisting of material objects, is the most important. This type of evidence speaks for itself in a courtroom, and it speaks with a loud and clear voice. It is very difficult to dispute what jurors can see with their own eyes, and feel with their own hands.

Circumstantial evidence, on the other hand, is sometimes difficult and complicated for the jury. The proof of various facts or circumstances, which either dispute the main fact of the case and prove its existence, or sustain by their consistency the hypothesis claimed, is hard for a jury to relate to. Circumstantial evidence consists of reasoning from facts, which are known or proved, to establish such evidence that is conjectured to exist. Physical evidence is easy for a jury to understand. The entire effort of the prosecution is trying to tie the evidence into the activities of the suspect. This is usually much easier to do with physical evidence than with circumstantial evidence. For this reason, protection of the crime scene is a critical part of any investigation. It is the main and usually the first source of physical evidence.

Physical Evidence

Physical evidence can consist of almost anything that has body or material substance. Fingerprints, weapons, empty cartridge cases, bloody knives, footprints, and matchbook folders are the usual things portrayed on television crime shows as being evidence. They are almost always quickly spotted by TV sleuths. In real life, however, this is seldom true. Searching for usable evidence at a crime scene is a painstaking job requiring knowledge and specialized training. The person looking for evidence must know what to look for and how to preserve it once it is found.

Physical evidence can be very difficult to locate and handle. Sometimes it can be difficult to preserve, even under the most desirable conditions. If it has been disturbed, kicked, or walked on, it may be destroyed as far as usefulness is concerned. This abuse of evidence is usually brought about by careless, nosy, or thoughtless intruders. Physical evidence at the scene of a crime must be protected. An entire investigation can depend upon its preservation.

Give some thought to the following evidence and, because of its nature or construction, how easily it could be destroyed by thoughtless tramping through the crime scene. Footprints can be left on a rug or in the dirt outside a window. A hair can be found on the floor, on a rug, or on a sofa. A burnt match can be found under the cushion of a chair or under a table. Grains of sand, pieces of soil, blades of grass, or other minute particles can be found in trouser cuffs. All these things, plus thousands of other small, inconspicuous items may be left by a criminal at the crime scene. Each object or particle could be critical in the solution of a crime, but because of their size they may be difficult to see. They are very easily destroyed by the careless trespass of an unthinking person.

WHY PROTECTION OF A CRIME SCENE IS NECESSARY

For the reasons mentioned above, protection of a crime scene is of the utmost importance. The more people who trample through the area where a crime has taken place, the more chance there is for contamination of the scene. The area must be closed off to all and protected from disturbance until the time the evidence specialists arrive and complete their work.

The Security Officer and the Crime Scene

It may not be often that security persons will find themselves in a position in which they will have to protect a crime scene. Should the situation arise, they should know how to handle it.

Should a security guard be confronted with an instance where he or she discovers a major felony crime has been committed within the confines of the area one is assigned to protect, the officer should stop immediately, survey the situation, determine the facts and problems as best as possible in the shortest span of time, and assign priorities to the things needed to be done. *To rush into the area without taking the necessary precautions and knowing what to do, and how to do it, could be a complete disaster.*

The first thing should be to determine whether a victim is involved in the crime. If it can be ascertained there is a victim, the victim's well-being is the first priority. Once the presence of a victim is confirmed, enter the area with all possible caution, being careful not to step on any visible objects, or to disturb or kick anything. Approach the victim in the most direct path possible, making note of where you walk. You will want to exit over the exact same route.

Determine whether the victim is alive or dead. If death is obvious, immediately exit the way you entered. If there is a question whether or not the victim is alive, check for a pulse on the wrist or at the carotid artery in the neck. One or other location is usually available to the touch without having to move any part of the body. If the victim is dead, do not move the body or disturb anything on the body, including clothing. Once again, upon determination of there being no life, exit the way you entered, call the police, and seal off the area until police arrive.

If you must use a phone within the crime scene area, carefully pick up the instrument, using thumb and middle finger of the hand, at the receiver end of the phone, as close to the earpiece as is possible. There is less apt to be fingerprints in this area than on the normal grip area of the phone. Note the exact spot where you gripped the phone so you are able to indicate the spot to the police. Do not use a cloth because that may smudge usable prints. If there is another phone near the scene, use it in preference to the one at the crime scene.

If the victim is in need of medical aid, approach the person in the same manner as outlined before, and determine the extent of the victim's injuries. If the injuries are minor and you are able to treat them on a temporary basis, do so, or call for the ambulance and the paramedics at the same time you call for the police. Return to the victim and remain there. Keep the victim from moving around, explaining about the possible destruction of evidence, until the police and the paramedics arrive.

If the medics arrive first, caution them to enter and leave by the same path you took, and to move about as little as possible in the performance of their duties.

Should it be that there is no victim, do not enter the area, but seal it off and contact the police.

If the area is small, such as a single room, secure the entrance and remain there to prevent intrusion until the police arrive.

If the area is larger, or outdoors, you may have to enlist the aid of others, if available, to assist you. Should this be necessary, explain to your helpers the reason for what you are asking them to do. In most cases they will be glad to help. If the area is outdoors, try to find sufficient material to rope off the area, using makeshift signs to identify it as a crime scene.

Allow No Entrance to the Crime Scene

Do not permit anyone to enter the area, regardless of who they may be, until the police arrive and take over the responsibility from you. The only exception, other than the first police officer to arrive, would be any requested medical personnel.

When restricting entry to the crime scene, *there should be no exceptions.* Even if the property owner, husband, wife, or the President of the United States wants to enter, they should be kept out. When the police arrive, the decision as to who may enter will be theirs. You will have done your job.

Use Common Sense

Common sense has to enter crime-scene protection the same as any other activity of a security officer. If an injured person, or a corpse, is in a building or area that is in danger of being destroyed by fire or explosion, you must use good judgment and remove the injured party or the corpse to protect or preserve either one. Under such emergencies, contamination of evidence takes a back seat. Some evidence is better than no evidence.

Remember, being the first person on the scene, you have a responsibility to protect the evidence within the area to the best of your ability. All it takes is common sense and sound judgment. Common sense will tell you that the more people who enter the crime scene, the more chance there is for contamination of the evidence. Do as you are supposed to do and the police and the prosecution will thank you; go charging into the area like a bull in a china shop and they will not look at you so favorably.

Who Is the Judge?

Keep in mind, evidence comes in all forms and shapes, and can be the thing you least expect to be of value. Do not take it upon yourself to make a judgment as to what is evidence and what is not evidence. Take no chances; consider everything as evidence and keep as far away from it as is possible.

Do not be frightened by the magnitude of your responsibility. It is not as threatening as you may think. If you take positive action you will do some good. The degree of good will depend upon how much you think out the actions you are to take, and how well you perform your task.

CHAPTER 13 PROBLEMS

PROBLEM 1

Officer James Henry is on night duty at the Zephyr Tool Company. It's been a lonely night and James wishes it was over. Things are so quiet, and on nights like this it seems his tour of duty will never end.

Jim completes his tour of the shop and heads back toward the office. He'd been in the office about an hour earlier and there was nothing going on in there either.

As Jim opens the door from the shop to the office he smells smoke and sees Mr. Adams, the shop owner, on the floor by the copying machine lying in a pool of blood. A revolver is on the floor about three feet from Mr. Adams' body. Fire has already reached the drapes and it looks as though in another few minutes the entire office will be on fire.

What should Officer Henry do?

CHAPTER FOURTEEN

Negligent Acts, Errors, and Omissions

From out of the old English Common Law has come a historic civil liability for negligent acts of an individual. The private security sector is no exception to the general rule.

TO WHOM DO WE OWE A DUTY?

Negligence is the failure to act as a reasonably prudent person would act, resulting in damage or injury to an individual to whom we owe a duty to act.

Fay Q. Hansen v *Skate Ranch Inc.*, New Mexico Ct.App.p2d 517,1982:

Floor guard at skating rink who is off duty, but acting within the scope of employment, subjects his employer to liability for his negligence.

Grand Food Inc. v *Geraldine Sherry, et al.*, Md.Ct.App.444,A2d,483,1982:

Jury awards bystander damages from guard's employer for mental distress caused by gunshot fired by guard chasing felon. Guard fired shot negligently.

The word "duty," as used in this section, relates to the duty or requirement that people shall conduct themselves in a particular manner. This requirement does not obligate or require anything of the person that is beyond one's ability to do or perform, since it refers only to the person's conduct over which one alone has control and is able to do.

The breach of a duty, or the conduct of a person that is contrary to the way one should act or behave, does not automatically make that person liable or accountable to others for one's actions. It merely subjects the person to liability. This means people may be held accountable for their actions. Whether or not one is liable and can be held accountable depends upon whether one's actions result in injury to someone to whom a duty is owed.

DUTY

The duty to do no wrong is a legal duty. The duty to protect against wrong is a moral obligation only that is not recognized or enforced by law.

Under our system of Common Law there is no legal duty for an individual to respond to another individual in need of help.

In deciding whether there is a duty owed by one person to another, there is present in much of the law a definite distinction or difference between the action and inaction on the part of the person doing or not doing the action.

In the early Common Law of our country, one who injured another by his or her act was held liable to the other person for the act without much concern as to whether or not the person was as fault. The courts were more concerned with the more obvious and flagrant forms of misbehavior on the part of the person performing the act than with someone who merely did nothing, even though another person may have suffered because of one person's failure to act. The law was reluctant to make the courts an agency to force people to help others.

Liability for failure to act was therefore slow to be recognized by the law. It first appeared in the law in instances of persons engaged in "public" callings, or jobs. These are individuals who, by holding themselves as servants of the public, such as bus drivers, police officers, firefighters, and public utility employees, serve the general public. In these and similar capacities they are regarded as having undertaken a duty they are committed to. They are liable for the results of their failure to perform certain duties. This idea of the law is still in effect today.

A bus driver owes a duty to assist passengers in distress.

The captain of a ship is required by maritime law to attempt to rescue a crew member who has fallen overboard, or to attend to an injured seaman.

The police officer has a duty, arising out of his or her occupation, to come to the aid of someone in distress.

Following the development of action to hold public servants liable for failure to perform, this principle was extended to anyone who, for a consideration or pay, had undertaken to perform a promise or act according to a contract. In cases such as *Henry Harris et al* v *Pizza Hut of Louisiana Inc., and St. Paul Fire and Marine Insurance Co.* (S.Ct.La.455, So.2d 1364,1984) the court ruled that a restaurant which employs a security guard to protect premises and patrons is liable for

> injuries due to guard's negligence in performing his (or her) duties.

In *Deeds* v *American Security et al.*, Ct.App.Ohio,39 Oh.App.3rd 31,1987:

> Private security guard and employer under contract to protect apartment complex only are not liable for personal injury to apartment complex guest. (This case emphasized general rule that a person does not have an affirmative duty to act to protect another. This case explicitly limited officer's duty to guard against vandalism or injury to the property.)

The extension of the principle has continued, and will most likely continue in the future. For the most part, the extension of a duty has been imposed on those who have, or have the potential to have, an economic gain from their position or contract. The largest group of persons so affected are owners or occupiers of land or real property.

LIABILITY FOR ACTIONS

Liability for actions taken against another may extend to any person to whom harm may reasonably be anticipated, as a result of the defendant's conduct, or perhaps even beyond.

Phyllis Tucker v *Ellis Smit d/b/a Sterling Secret Service Inc.*, and *Robert Rothget* v *Stewart Mott Community College*, Mich. Ct.App.337 N.W.2d 637,1983:

> Security officer was found liable for failing to notify police of the presence of an assailant thereby permitting assault on a college student.

See also *Dina Cruz, individually and a parent and guardian of infant plaintiff Moises Cruse* v *Madison Detective Bureau Inc.*, v *RKO Theaters* for the following ruling:

> Movie theater usher assaulted by patrons can maintain an action against a contract security officer for failing to provide adequate security at theater and failing to follow its own security procedures.

Liability for not taking action requires the finding of some definite relation between the parties of such a character that social policy justifies the imposition of a duty to act.

Failure of an engineer to blow a train whistle at a road crossing is negligent operation of a train.

Failure to repair a gas pipe is regarded as negligent distribution of gas.

One who digs a hole in the middle of a public highway has a duty to put lights or other warning devices around the hole so, as day approaches, users of the highway will be warned of the danger caused by the existing hole or excavation. Failure on the part of the person who dug the hole to put out adequate warning devices is a breach of that person's duty, but the mere fact of not placing the warning devices in place, of itself, does not make the person liable. The person does not become liable until someone, not seeing the hole falls into it and is injured.

DUTY OF CARE

A person who acts or does something must exercise reasonable care to make one's actions safe for others. *Safeway Stores Inc.* v *George I. Kelly,* D.C. Ct.App.448A 2d 856,1984:

Store which uses contract security guards but controls their conduct is liable for guard's excessive use of force in lawful arrest of customer.

Gulf Oil Corp., Empire Security Services, and Empire Security Agency Inc. v *Thomas Williams,* Tex.Ct.App.642 W.W.2d.270.1982:

Guard's activities were jointly controlled by both contract security company and the retailer. Both were held liable for guard's mistake in shooting customer.

The keeping of wild or ferocious domestic animals, and the carrying on of abnormally dangerous activities such as dynamite blasting, are typical situations in which a person's conduct is at the risk of having to answer for any harm these activities cause although individuals may use the utmost care in caring for the animals or in preparations made for the blasting, they still are at risk. However, in neither of these instances is the conduct of the person involved morally or socially wrong.

KINDS OF NEGLIGENCE AND PROXIMATE CAUSE

Negligence can be either active negligence—that is, negligence in doing an act (errors), or passive—negligence as a result of failing to perform an act (omission).

In order to be liable or held to account for an act of negligence, active or passive, *the negligence must be the proximate cause of the damage or injury to the person to whom a duty is owed.*

If an employee commits a negligent act in the course of his or her employment, the negligent act creates a civil liability not only on the employee but also on the employer.

If a security guard while on duty negligently causes a fire that damages the property of the client, the negligent act creates a liability for both the guard and the guard's employer.

In the case above, the guard committed a negligent act during the course of his employment, causing damage to the property of one to whom he owed a duty, namely the client. His negligent act was the proximate cause of the client's loss.

A less clearly defined liability may exist where the guard falls asleep while on post and fails to detect a fire caused by a short circuit in the electric wiring. The issue in this situation is whether the passive negligence of the guard falling asleep (omission) was the proximate cause of the damage to the client's property.

Early courts in most jurisdictions would have found the damage was caused by the short circuit and not the sleeping security guard. However, more recently there has been a definite trend to find that part of one's duties is to detect such emergencies. Failure to detect a fire because of being asleep when one is supposed to be awake was a proximate cause of a major portion of the client's loss. Courts would lean to the view that, although the guard could not have prevented the short circuit, he or she could have detected the fire close to the time it started, turned in the alarm, and thereby greatly reduced the amount of damage.

Ever since the early part of the century, most jurisdictions of the country have consistently held that the failure of a security service or system could not be liable for the criminal acts of a third person. In these cases, the courts have regularly found the proximate cause of the loss to a client was the result of the criminal activity of the third person and not the negligence of the security service.

GREATER DEGREES OF CARE

Innkeepers (motels and hotels) and landlords, at Common Law, have always owed a greater degree of care to their guests and tenants than would be found in ordinary cases. Courts have used this strict innkeeper and landlord liability to hold such individuals liable for injuries to guests and tenants caused by the criminal activities of third parties for negligently failing to provide adequate security and safety systems.

Larry Kranz and Joynce Kranz v *La Quinta Motor Inn Inc.*, La.S.Ct.410 So2d.1048,1982:

Sleeping guard and negligent delivery of pass key render motel liable to guests for third party assault, battery, and theft.

Jackie Berniece Boles v *La Quinta Motor Inn Inc.*, Tex.U.S.Ct.App.5th Dist.680 F2d 1077,1982:

Hotel not liable for causing rape of guest is nevertheless liable for aggravation of injuries caused by delay in aiding victim.

Thomas H. Urbano v *Days Inn of America*, N.C.Ct.App.295 S.E.2d 240:

Hotel may be liable for failure to warn guest and to prevent criminal assault where prior criminal activity in area should have made attack on motel guest forseeable to owner/operator.

GOOD SAMARITAN LAWS

Good Samartan laws are designed to protect persons, most often doctors, from suits arising from the rendering of free assistance in an emergency. California enacted the first such statute in 1959. Since then more than 43 states have passed similar laws.

None of the Good Samaritan statutes impose a duty to render aid; they only insulate those who render aid from liability for ordinary negligence.

The law has persistently refused to recognize the moral obligation of common decency and common humanity to come to the aid of another human being who is in danger.

An expert swimmer who has a boat and rope nearby and who sees another person drowning is not required to do anything about saving the person. That individual can stand on a dock and watch the other person drown.

A physician is under no duty to answer the call of one who is dying and could possibly be saved.

A person is not required to bind the wounds of a stranger bleeding to death.

The remedy in cases such as these is left to a person's conscience. Decisions of individuals not to help in such cases are revolting to any moral sense and have been denounced vigorously by society.

REASONABLE CAUSE

In instances where a duty to another or others is recognized, it is usually agreed that the duty calls for nothing more than reasonable care under the circumstances. A person is not liable when he or she neither knows, nor should know, of the unreasonable risk, or of the illness or injury involved. One is not required to provide aid to someone whom one has no reason to know to be ill. People will seldom be required to do more than give such first aid as they reasonably can, and take reasonable steps to turn a sick or injured person over to a physician, or to those who will look after the sick or injured person until a physician can be contacted.

LIABLE FOR MAKING SITUATION WORSE

If there is no duty to come to the assistance of a person in difficulty or serious danger, there is at least a duty to avoid any affirmative act that will make the person's situation worse.

If someone does undertake rescue efforts, he or she may be held liable if that individual performs them negligently. (see *Zalenko* v *Gimble Bros.,* 158 Misc.904,287,N.Y.S. 134, 1935.)

There may be no duty to take care of, say, a man who is ill or intoxicated and unable to look out for himself, but it is another thing entirely to eject him into a dangerous situation such as a busy street. If the man is injured there will be a liability.

In addition, if someone does attempt to aid the person mentioned above and takes charge and control of the situation, that individual is regarded as entering voluntarily into a relation that is attended with responsibility.

Again, it is important for the student to be reminded that each state has its own statutes and laws. The laws of your state may not conform to any of the examples contained here. Thus it is important that you check the laws of your own state to see how they apply to your situation.

CHAPTER 14 PROBLEMS

PROBLEM 1

Security Officer John Welsch is driving on Highway I-44 about midnight on Thursday. He was relieved at his assignment at 11:00 P.M., and is on his way home. The rain has stopped but the highway is still wet and somewhat slippery. John will be glad to get home tonight. It seems as though his shift was longer than usual. He is tired and doesn't feel up to par.

John notices that there isn't very much traffic on the highway at this time of night. He is glad because he doesn't have to face the oncoming headlights.

While rolling along at a moderate rate of speed John notices what appears to be a fresh set of skid marks breaking sharply to the right, and what appears to be freshly dug sod along the apron of the road. Slowing down, he looks into the ditch on the right side of the road and sees auto tail lights shining up at him from the hole. A vehicle is nose down in the trench and the horn is blowing.

It appears to John that an accident took place moments before he arrived. No one else seems to have stopped.

Should John stop? Does he have an obligation to do so? By doing so, does he assume any responsibility?

CHAPTER FIFTEEN

Giving Testimony and Courtroom Behavior

Security officers, having little opportunity to appear in the courts in the line of duty, do not often realize the importance of courtroom behavior and the proper way of giving testimony. The court appearance will be a very good test of officers' report writing skills, their work abilities, and their ability to maintain a clear mind and to act with intelligence when in a stressful situation. It may also challenge their knowledge of security, their individual character, and their personal lives. For the unaccustomed security officer this can be a very big moment.

It is nothing to be afraid of but it is something you must be prepared for.

Security personnel who work in retail merchandise establishments have more opportunity for court appearances. They are therefore more familiar with the entire proceedings. If they stay in the retail field of security very long they usually become proficient witnesses.

The best written report and the most capable prosecutor cannot overcome a poor impression left on the court, or jury, by an unqualified, poorly prepared witness.

To make the inexperienced security person comfortable with this new experience and the thought of having to appear in court and give testimony, it may help to provide a comparison to something that the officer may be more familiar with.

THE COURTROOM

Picture yourself going to see your first stage play. Never having seen a stage play, even though your friends may have told you about them, you are not sure what to expect when you walk into the theater. You have been told about this play or that play, but no one bothered to tell you about the appearance of the theater. It can be overwhelming. So too with your first courtroom appearance.

You walk through the swinging doors of the courtroom and take a seat in the rear. You are able to see the entire interior almost at a glance. You are not sure if it looks like you thought it would look.

For the most part, all courtrooms are quite similar. Some of the newer ones have more modern interiors, and older ones have a nostalgic look. Old or new, they have the same general floor plan and makeup.

The one thing that catches the attention of the first-time visitor is the highly elevated structure in the center of the courtroom. It is the pivot about which everything else seems to be built. Because it is the highest place or object in the room, it radiates authority. Rightfully so; this is the judge's bench. The judge's high-backed chair sits majestically behind the bench, looking down on everything and everyone below. Normally the judge's bench is very large, larger than any other object in the courtroom.

From this lofty position, the judge, dressed in judicial robes, is able to look down on all other persons and things in the room. Conversely, everyone must look up to the judge. This elevated central location in the courtroom leaves no doubt in anyone's mind about the position and authority of the person who occupies the massive chair. From this lofty position the judge has the duty of directing the flow of progress in a trial. The judge specifies the issues of law governing the case at hand and rules on any and all points of law that may arise during the proceedings. The judge is the lord of this room; it is the judge's domain.

The next attraction is the raised platform with a chair in the center. It is sometimes surrounded by an ornate railing; then again there may be no railing at all, simply a microphone. Though the platform is raised, it is not nearly as high as that of the judge's bench. However, it is usually the second highest point in the courtroom. It can be seen by everyone from any location inside the room. It is from here that witnesses give their testimony for everyone in the room to hear. All eyes come to rest on the person occupying this chair, for this is one of the most important persons in the proceeding. Without the witness there would be no trial. It is the testimony of the witnesses around which the trial is based.

At floor level, directly in front of, and at the base of the judge's bench, is the desk and chair of the court clerk. The court clerk has the duty of swearing in witnesses, identifying and marking prosecution and defense exhibits of evidence used during the trial, and keeping necessary records for the court.

Near the clerk's desk is located another desk or, at times, simply a steno's stand used by the court recorder. He or she records every word said by all participants in the trial. This is done for the entire length of the trial, including what is said by the attorneys to the judge, in front of the judge's desk out of earshot of the jury, or in the judge's chambers. These recorder's notes are so accurate that they are referred to and asked to be read back aloud to the court during the trial. These notes become the trial transcript and are used in the event of an appeal.

At floor level, on one side of the room or the other, depending upon where the jury box is located, are the desk and the chair of the court bailiff. In most instances the bailiff is a deputized officer. It is the bailiff's job to call the court to order, to maintain order in the courtroom at the direction of the judge, and to escort prisoners into the courtroom if they are presently in custody. The bailiff usually works in close conjunction with the deputy, who brings into the courtroom a defendant who has been locked up.

Directly opposite the bailiff's position is a group of 14 to 16 chairs. Twelve of these chairs are for the jurors, and the remainder are for the alternate jurors. In some courtrooms these chairs rest on an elevated platform and have a railing in front of the jurors. This is the jury box and is the place from which the jurors and alternates listen to the testimony as it is given during the trial. It is almost always located on the same side as the witness chair so jurors can clearly see and hear the witness giving testimony, and the witness can see the jurors.

In front of the judge's bench, and in front of the clerk and the recorder's desks, are two long tables that are separated from each other by at least a few feet. One table is for the prosecution team and the other is for the defense team.

In a great majority of courtrooms, a railing with swinging gates divides the room into the area in which the trial actually takes place and the area where spectators, who wish to observe the proceedings, are seated.

The room, the high-backed chair, the high-positioned bench, the jury chairs, and the ornate railings are impressive, to say the least. But this is when the room is empty. This seemingly majestic setting takes on a new life when the members of the court take their places and there is an indication the trial is about to start.

THE HIGHLY VISIBLE ACTORS

Near the beginning of this chapter we mentioned going to a stage play for the first time. Well, the court has its own cast of characters, each with a part to play. After all these cast members have played their part, the play (trial) comes to an end with either a happy or a sad ending.

We have discussed the parts most of the members of the cast play, but we did not mention the duties and responsibilities of the jurors, the prosecuting attorney, or the defense attorney.

It is the duty of the jurors, after having been chosen by the combined efforts of the prosecuting and defense attorneys, to listen attentively to all testimony given by all persons during the trial. After hearing all of the testimony and seeing all of the evidence (physical objects), the jurors are instructed by the judge on the issues of the law governing the case being tried. The judge's instructions are clear and explicit, and any questions the jurors may have during their deliberation can only be asked of, and answered by, the judge. The jurors weigh the facts of the testimony and the evidence and render a verdict for either the prosecution or the defense.

The prosecuting attorney represents the people of the state in which the incident on trial took place. It is his or her job to present to the jury the state's evidence in an honest, forthright manner. Prosecutors must make every effort to convince the jury, based on the evidence presented during the trial, that the defendant is guilty of the offense charged, beyond a reasonable doubt.

The defense attorney represents the defendant. This person has some of the same duties as the prosecuting attorney except his or her efforts are on behalf of the defendant. Defense attorneys have to convince the jury the defendant is innocent of the offense charged.

The defendant is the subject of the trial, since it is his or her alleged acts, or lack of action, which causes the court to convene.

The police officer, the detective, and or the security officer are aids to the prosecution or the state, since it is their reports of the alleged wrongdoing on the part of the defendant that form the basis upon which the state's case is built.

There is one very large difference between the courtroom drama and the stage play. When actors in a stage play forget or bungle their lines, there is a prompter in the wings to assist them in their presentation. Not so in the courtroom drama. Each and every actor, especially the witnesses, are totally on their own. There is no help available from anyone. Police officers or security officers who are witnesses especially must know their lines without hesitation. They must give their testimony with clarity, accuracy, and with certainty. Mistakes or bluffing on their part usually has an extremely negative effect on the judge and jury. Officers are subject to the same rules of truthful testimony as are all other witnesses.

The effect of the police officer's or possibly the security officer's testimony is based on a number of factors. Each is important, but all must be able to stand the rigid scrutiny of cross examination of a dedicated, knowledgeable, and determined defense attorney. Errors on the part of police or security personnel, revealed in cross examination, can be devastating to the state's case.

The courtroom is often called the "Hall of Justice." It is the place where justice is handed down, equally for all. The symbol of justice, a majestic blindfolded lady holding the scales of justice, becomes a reality in the courtroom. The blindfold does not mean Lady Justice is blind; rather, it means that by not being influenced by those she sees before her, she can dispense justice fairly to all people. Thus, who or what one is in life is not supposed to have an effect on Lady Justice. Though one side eventually wins and the other loses, the determining factor that tilts the lady's scales to one side or the other must be the evidence alone. To her, it is not important who wins or loses, only whether justice has been served.

Justice Does Have Some Weaknesses

It is no secret, although it is not supposed to happen, that human nature, personal egos, and at times politics find their way into a trial. There is no such thing as a perfect world, or a perfect society. These things happen, but when they do, immediately upon discovery, they should be eliminated. The stakes at a trial, especially a criminal trial, are entirely too high. Here we are dealing with the future freedom or reputation of another person. Such high stakes should never be the ransom to satisfy the desires or wishes of any individual, organization, or special interest.

The Obligation of Security Officers During a Trial

A very important point for all police and security officers to remember is that it is not the job of officers to try to obtain a conviction. Their function is to present unbiased and truthful facts, as they know them, in testimony. It is not their job to judge anyone. The conviction or acquittal of the defendant is the job of the jury. It is the jury's function to decide this based solely on the facts as presented to the jurors in court by the prosecution and the defense. Police and security officers may have formed an opinion based on what evidence they may be aware of, but this opinion must not be expressed in any way. It cannot determine or effect the manner in which their testimony is presented to the court. These officers may know only a few of the facts, and a knowingly biased presentation will live with them the rest of their lives. They could also be charged with perjury.

EVIDENCE USED IN A TRIAL

There are many important and critical elements of a trial. Of these, evidence is the most critical. Evidence is so important in deciding the fate or fortune of a person on trial that our justice system has set down a canon of rules governing it. These rules of evidence are so critical it behooves every police and security officer to know all there is about them. It would be unthinkable for an officer to make a mistake that could deny the prosecution the right to admit evidence that may cause a complete dismissal of charges against a guilty party.

Before discussing the different types of evidence, it is best to know the meaning of the word "evidence."

Black's Law Dictionary defines evidence as "any species of proof, or probative matter, legally presented at the trial of an issue, by the act of the parties and through the medium of witnesses, records, documents, concrete objects, etc., for the purpose of inducing belief in the minds of the court or the jury as to their contention." (*Hotchkiss* v *Newton,* 10 Ga.567; *O'Brien* v *State,* 69 Neb.691,96 N.W.650, *Hubbell* v *U.S.,* 15Ct.Cl.606).

Evidence includes all means by which any alleged matter of fact is submitted to investigation and is established or disproved (*Dibble* v *Dimic,* 143 N.Y.549,554,38 N.E.724,725).

The rules of evidence are critical to a trial, and though they are interpreted for the state and the defense by the judge, it is important for police and security officers to have a sound knowledge of these rules. This will assist officers in not making a mistake during the preliminary stages of an investigation that could negate the entire case.

Physical Evidence

1. Physical evidence can be either real or demonstrative, or documentary.
   a. Real or demonstrative evidence could be a weapon, a fingerprint, a footprint, blood, clothing, etc.

   (Real evidence is that evidence furnished by things themselves on view or inspection, as distinguished from a description of them by the mouth of a witness.) *Black's Law Dictionary,* 4th edition, 1968

   b. Documentary evidence can be a report, bill of sale, birth certificate, marriage license, etc.

Circumstantial Evidence

Circumstantial evidence consists of facts or circumstances from which the existence or nonexistence of fact in issue may be inferred. (*People* v *Steele,* 37 N.Y.S.2d,199,200,179 Misc.587). Circumstantial evidence consists of inferences drawn from facts proved. (*Hatfield* v *Levy Bros.,* 18 Cal.2d,798,117P 2d,841,845). Circumstantial evidence is a process of decision by which court or jury may reason from circumstances, known or proved, to establish the inference to the principal fact. (*People* v *Taddio,*292 N.Y.488,55,N.E.2d 749,750). Circumstantial evidence consists of theoretical facts, not necessarily substantiated by actual facts, that tend to lead to an unproven conclusion.

Hearsay Evidence

Hearsay evidence is evidence not preceding from the personal knowledge of the witness, but from the mere repetition of what the witness has heard others say. It is evidence that does not derive its value solely from the credit of the witness, but rests mainly on the veracity and competency of other persons. The nature of the evidence shows its weakness, and it is admitted only in specified cases, from necessity. (*State* v *Ah Lee,* 18 Or.540,23P,424,425; *Young* v *Stewart,* 191 N.C.297,131,S.E.735,737.)

> Hearsay is secondhand evidence, as distinguished from original evidence; it is the repetition at secondhand what would be original evidence if given by the person who originally made the statement. Literally, it is what the witness says he heard another person say. (*Stockton* v *Williams,* 1 Doug.Mich., 546,570)

To be admissible, evidence must meet the following definitions as found in *Black's Law Dictionary.*

1. Relevant Evidence—"That which relates to or bears directly on the point or fact in issue, and is admissible."
2. Material Evidence—"That which is relevant and competent, and goes to prove the substantial matters in dispute, having a direct, legitimate, and effective influence on the decision of the case."
3. Competent Evidence—"That which is relevant and applicable to the case, and is capable or competent to present it."

Complete and accurate notes have often been responsible for the success of a prosecution. Although the usual use and purpose of notes is to prepare reports, officers should not forget that their notes may be used in court on occasion as an aid to testifying.

THE USE OF NOTES IN TESTIFYING

In many instances, especially in the career of police or security officers, a considerable amount of time may elapse before a case in which they are involved comes to trial. In the interim, these officers most likely handled a number of cases, and taken a great number of notes. For this reason, it may be difficult for them to remember all of the specifics of the case at trial, and they may need to refer to the notes they had taken in preparation for the report they wrote concerning the case.

Both Common Law and statutory law of some states provide that the memory of officers may be refreshed by the use of these memoranda. When a police officer or security officer witness, after obtaining the permission of the judge, reviews reports he or she prepared at the time of the case and still cannot recall facts in sufficient detail, and the officer testifies that he or she once knew the facts but does not recall them enough to giver proper testimony, notes in the officer's own handwriting taken at the time may be received in evidence, although the officer may have no present recollection of them.

McCarthy v *Meaney,* 183 N.Y.190,76 N.Y.S.36
Conn.State v *Masse,* 1 Conn.Cir.381, 186,A.2d 553

If the witness uses a memorandum to refresh his or her recollection, even though it may not be shown to the jury as evidence, it may be used by opposing counsel in that counsel has a right to inspect it and use it to test the credibility of the witness.

People v *Gezzo,* 307 N.Y.385,121N.E.2d 380

If two officers, be they both police officers, a police officer and a security officer, or two security officers, work together on a case, they should read each other's notes and initial each page. Either officer can then testify from the notes.

Once again, the reader is cautioned that each state has its own laws and rules of procedure for its courts. The rules in your state may be different from the rules stated above. Each person should double-check the rules of his and her own state concerning courtroom procedure.

DISCOVERY PROCEEDINGS

The discovery proceeding is a well-established legal right of the defense to have access to all evidence materials that the prosecution has. When presented in court, the prosecution must allow the defense the opportunity to examine notes, documents, or other memoranda.

Under discovery proceedings, the defense is permitted to request any notes of conversations between the defendant and security officers taking place prior to trial.

COURTROOM APPEARANCE AND BEHAVIOR

One of the factors that may have an effect on the court, and especially on the jurors, is the outward appearance of an officer giving testimony. It is known that jurors place a great deal of weight and emphasis on witnesses' appearance. Some are of the opinion that laypeople believe a neat-appearing officer is more readily believable than one who comes to court in disarray.

Officers, be they police or security, should by their actions and behavior appear to be completely neutral. Because of the officer's uniform, the slightest indiscretion in behavior tends to give jurors the opinion the officer's testimony will lean toward the prosecution. Although the prosecutor may speak to the officer before the officer actually takes the stand, the officer's conduct during such a conversation should give the appearance of only speaking when spoken to. There should be no indication of lightheartedness or smiling on your part. Neither should you walk around with a mean or officious look.

Often, in order to help keep jurors from feeling the officer may not be totally neutral, the prosecutor will request that an officer wear civilian clothes to the trial. Prosecutors want to remove all chance of the uniform putting wrong ideas into the mind of the jury.

When the officer comes to court in civilian dress, any sight of a weapon, handcuffs, or even a bulge under a coat, indicating the presence of such items, will negate the entire purpose of wearing civilian clothes.

Regardless of the number of times an officer has been called upon to testify in court, there is always some trace of nervousness or uneasiness. Some cases may have a more emotional effect on the person having to testify than another. Because of this, an officer should make psychological and mental preparations ahead of time. If an officer is able to walk calmly into a courtroom, take the oath, take a comfortable position in the witness chair, and give firm, straightforward testimony in a strong voice, jurors will be impressed. They will see, and have the opinion, the person on the stand is a professional doing his or her job in a like manner.

To say a witness, no matter how professional, will never experience some nervousness is not true. The nervousness may not show on the outside, but for a few seconds, until actually seated on the witness stand, everyone feels a bit uneasy. This is because of the ever-present fear of the unknown. Even though the officer feels confident he or she has all of the answers, there is a second or two when doubt enters one's mind; once again, this is because of the unknown. Attorneys all have their own style of examination, and until the style is revealed to the witness, there is that unknown feeling again. It is perfectly normal. First-time witnesses believe it only happens to them. They are wrong. It happens to everyone.

Always remember, from the time your name is called to come forward to be sworn, until the moment you begin your testimony, the eyes of all the jurors are concentrated on you. They are sizing up your appearance, your demeanor, and your professionalism.

To eliminate the possibility of any one or more of the jurors taking issue with you as an individual, do not wear lapel pins signifying membership in any organization or affiliation. Do not wear tie pins or clasps that give like appearance. There may be those on the jury who are opposed to the affiliation you make known. Such displays may prejudice their thinking against the prosecution's case.

Before being called to the stand, be it in a hallway, a waiting room, or the courtroom, it is always good for an officer to go over a mental checklist of important points to remember:

1. Are my *original notes* available? Are they in good condition so they may be entered into evidence and stand the scrutiny of the defense?
2. Have I familiarized myself with any lab reports, reports on the defendant, or other reports which have a bearing on the case at trial?
3. Am I aware of the witnesses to be called? Is there anything particular I should know about them?
4. Have I conferred with my supervisor, watch commander, or any other necessary supervisor?
5. Have I discussed the case with the prosecutor so he or she is familiar with what my testimony will be?
6. Is my appearance as best as I can make it?

An officer's preparation for going to trial is always obvious to the court and the jury. Even though jurors are not sure why they feel the officer is ready, they have the impression it is a well-trained, disciplined person who is speaking from the witness stand.

An officer should appear bright and pleasant, without overdoing it. After all, there is most likely not much joy in the heart of the defendant. To give jurors the impression you are truly enjoying testifying against the defendant will give them a negative impression of you. It is not your job to convict the defendant, but simply to present your evidence, through the prosecutor, in an outward and unbiased manner.

Politeness is a common trait, even outside the courtroom. Inside the courtroom it can be very helpful; lack of it can be destructive in the mind of a juror. It can almost be inferred with certainty that the witness a juror likes or feels good about is bound to receive the benefit of the doubt on any border-line question or answer. This is not to say witnesses are supposed to be actors. This type witness can usually be seen through easily. What it means is that witnesses should be themselves, be polite and considerate, show high respect for all of the court officers and lawyers, and especially members of the jury. These dozen people, almost subconsciously, desire to be recognized for the task before them. Many have sacrificed a lot personally to be there. To antagonize them or lose their confidence is a terrible loss to the prosecution's case.

Remember also that politeness carries far beyond the courtroom itself. An officer who is to be a witness may enter an elevator on the way to court. The elevator could be crowded, and among the crowd may be one or two people reporting for jury duty. A polite and courteous officer, holding the elevator door open for others on the elevator to exit first, may just have won an important vote of confidence. These people may be chosen for, or already be on, the jury in front of whom the officer will later testify. What goes through the mind of the officer's elevator companions? Could it be something like, "There's that nice officer who held the door open for us"? Their first impression of the officer is carried with them into the jury box without them ever knowing the officer's actions would have an effect on them so soon.

There is nothing more embarrassing for a prosecutor than to have to explain to the judge why an officer witness is late in arriving to give testimony. The judge may permit the calling of a different witness, but you can be sure he or she won't like it and will have some harsh words for the officer in chambers, when the officer does arrive. Whatever makes the judge resentful as far as witnesses are concerned usually carries over to the jurors.

Upon arriving at court, officers should notify the prosecutor of their presence.

It may have been some time since you last saw the defendant. Many cosmetic or physical changes could have taken place during that period. Hair could have been trimmed or grown, a mustache or beard could have been grown or removed, or accidents or fights could have caused physical changes in the person. Locate the defendant so you remember as much about the person as is possible. If you cannot recognize the defendant for any reason, have the prosecutor point him or her out to you if there is time and only if this is not done in the courtroom. It would be embarrassing and damaging to the prosecution if you were asked under oath to point out the defendant and were unable to do so. If you are not able to get the information about the defendant from the prosecutor, for one reason or another, locate the defense table and attorney. Look at the people around the defense attorney or those seated at the table. The defendant is seldom more than a few feet from the defense attorney unless, of course, the attorney is in the judge's chambers.

Some jurisdictions allow called witnesses to take a seat in the spectator's section of the courtroom until summoned to the stand; other jurisdictions do not permit this. In most jurisdictions the witnesses must remain out of the courtroom during the testimony of others. The main reason for excluding future witnesses in the courtroom before they testify is the fear that the testimony of other witnesses will possibly effect the way they will testify.

If you are permitted into the courtroom, sit erect, be alert, pay attention to the proceedings, and when you are called to the stand to be sworn in, advance in a professional manner. Raise your hand to take the oath in a firm, respectful manner, and speak distinctly and loud enough for all in the courtroom to hear.

Take your seat in the witness stand and keep both feet on the floor squarely in front of you. Face your seat more on an angle toward the jury than to the front of the court. Look directly at the attorney who is questioning you, but when answering his or her question, split the difference between looking at the attorney and looking directly at the jury. In this manner, jurors will certainly hear your answer as will the attorney. Speak clearly, loudly, and firmly. Do not act hesitant or fearful that you will not say the right thing. *You are sworn to tell the truth, and the whole truth.* Whether you or anyone else likes it or not, you have no other choice. Relate the facts the way they happened with nothing added and nothing left out.

THE ACTUAL QUESTIONING

Under ordinary circumstances, the prosecutor will begin the examination of the witnesses. Under what is known as "direct examination" no leading questions can be asked of you. Direct examination during the prosecution's half of the trial belongs to the prosecuting attorney. During the defense's half of the trial, direct examination belongs to the defense attorney.

A leading questions may be, as an example, "Was the car blue?" The proper way of asking this question during direct examination would be, "What was the color of the car?"

Usually the first questions asked of you will be of the type required to identify yourself, your place of residence, your occupation, your rank, your present assignment, and perhaps even your educational background, including specialized areas of expertise. During the time you are answering these and other future questions, try constantly to be aware of the tone of your voice and the attitude with which you are answering the questions. Your voice should be loud enough for all to hear, totally free of arrogance or superiority. Your speech should be evenly paced, not fast, not slow. Your pronunciation should be as accurate as possible and not in the least overbearing. The inflections in your voice should give no indication of any leanings to one side or the other you may have. If you have such leanings, keep them entirely to yourself. Do not talk *at* the jurors; talk *to* the jurors. Do not give them the impression you are trying to force your testimony on them or that what you have to say is too complicated for them to understand.

Your testimony will be evaluated and weighed by both the judge and the jury. You will be under very close scrutiny the entire time you are on the witness stand. Your attitude should not change a single bit whether the questions are coming from the prosecutor or from the defense attorney. Wait until the question is asked, and answer only the question you are asked. Do not volunteer anything that is not requested of you. The defense attorney will be listening intently and will be able to cross examine you on any statement you volunteer. Without knowing it, by volunteering and opening the door to a subject, you may be hurting the prosecution, whose witness you are.

The prosecuting attorney will control the direction of the case by the questions he or she asks. Prosecutors bring out the testimony they want brought out by their manner of questioning. If you think it is strange the prosecutor does not ask certain questions, be sure that he or she has a reason for not asking. Do not volunteer anything; you may be saying just the thing the prosecutor doesn't want to hear. Prosecuting attorneys know well what they are doing. They don't need your help. They have spent weeks and months of time preparing their case and plan of presentation. You are the last person to complicate things. Trust the prosecuting attorney. It's his or her case to win or lose. Let the prosecutor do it on his or her own. If you do things the way you are told and answer questions the way the prosecutor wants, then you are a part of the team that helped win the case. If you damage the case in any way you will always be remembered as the officer who helped the prosecutor go down to defeat.

Be sure any question asked of you is clear and you understand it. If you are not certain of the meaning of the question the way it is phrased, state the fact you don't understand the question and request it be restated in another manner or that it be clarified.

Do not be anxious to answer questions rapidly. There are two reasons for this. First of all, it can give the appearance you may be a know-it-all. The second is to give the question adequate thought and to allow the defense attorney enough time to offer any objection he or she may have to the question. The judge will rule on the objection, and according to the ruling you will know how to proceed. If you answer a question before an objection can be raised or the judge is able to rule on the objection, the question and answer will be most likely stricken from the record, and the jury will be instructed to disregard both the question and the answer. When this happens too often with the same witness, not only is the attorney angered but the judge may also show outward signs of irritation. This can possibly have a negative effect on the jury.

Be polite and courteous in your answers. Use "yes, sir" or "no, sir" when speaking to either attorney and "your honor" when speaking to the judge. Do not try to impress anyone in the courtroom by using police or security jargon. Do not use vulgarity unless you are quoting from a statement of another. If this is necessary, offer an apology to the judge and the jury for having to use the vulgarity. Let them know you are not speaking in such a manner by choice.

When answering questions, look at the counsel addressing you while he or she is speaking. Look at the jury when giving your answer. Direct your attention to the center of the jury box when you are speaking to jurors. Movement of your eyes will allow each juror to feel as though you are speaking directly to him or her. When you have stated your answer to the question, stop. A favorite ruse of the defense attorney may be to stare at you as if expecting you to continue. Do not get tricked into this situation. Look directly at the defense attorney. Display a relaxed attitude and await the next question. Do not ever show displeasure with the defense attorney by your actions or your answers. Defense attorneys represent clients. Just as the prosecutor, the defense attorney will use all the techniques and knowledge he or she has to win the case. There is nothing personal in the questions you are asked, or in the manner in which they are asked.

If your testimony happens to favor the defendant, but is true, do not hesitate. The defendant is entitled to a fair trial. He or she cannot get it if the testimony of witnesses is deliberately false. This can result in a charge of perjury against the witness.

RULES OF EVIDENCE

Earlier we spoke of some basic procedures concerning the rules of evidence. You must be aware of the basics, and what is admissible and what is not. To be clever and slip in something that is not supposed to be there is to ask for a mistrial. And if not a mistrial, at least grounds for appeal. A major point to remember is not to make any statements concerning an inadmissible prior conviction, or the prior conduct of the defendant. Errors such as these are also subject to challenge from the defense and can be grounds for a mistrial or can open the gate for the defense to a whole new strategy of questions.

An inquiry may be made of you as to whether you have discussed your testimony, prior to the trial, with the prosecutor. If asked this, do not deny it. The jury is probably aware of the fact the prosecutor has more than likely discussed with you your testimony. Answer affirmatively. This tells the defense you are aware of the attempt to try to trick you or to anger you. If you leave sufficient time between your answers and do not attempt to answer immediately, you will not be caught in this trap. This pause also allows the prosecution to object to insulting, embarrassing, or shaming questions put to you by the defense. It will also give the judge time to sustain the prosecutor's objection. Instead of embarrassing you, you will have turned the tables.

If at any time during your testimony you realize you have erred, turn to the judge and say, "Your honor, I just realized I made a mistake in my prior testimony and I ask permission to correct my testimony for the record." You will be better thought of by the court and the jury for determining your own mistake than having the opposition embarrass you by bringing it to light later in the trial.

If you do not discover your mistake until after you have left the witness stand, notify the prosecutor at once. In all likelihood he or she will recall you to the stand and allow you the opportunity to set the record straight.

While you are on the witness stand, avoid all eye contact with the prosecutor unless he or she is the one asking you a question. Do not give anyone in the court the idea you are looking to the prosecutor for assistance. You are on your own. To be constantly looking at the prosecutor will surely give the jury the wrong impression and will detract from the believability of your testimony.

Unless you know exact figures when answering questions regarding distance, height, weight, time, and so forth, use estimates. Never make an outright guess. If you must resort to that, it is better to say you don't know, or you are not sure.

Should you be called upon to put an illustration on a blackboard or point to something on a diagram, be sure you take a position that will not block the view of the jury, the judge, or the attorneys. If this cannot be accomplished from one place, make a first presentation, then move to another angle and repeat the presentation so all will be able to see.

There are times when a witnesses must testify regarding conversations they may have had with defendants. Certain rules must be followed in such cases. On direct examination the witness can testify only to the conversation he or she had with the defendant, or in some instances, a conversation the witness may have had in the presence of the defendant, where the defendant could hear and understand the conversation. In most cases the prosecution will lay a foundation before introducing a conversation. This is done by a number of lead-in questions, such as:

"Now, officer, at any time, did you have a conversation with the defendant?"
"Where did this conversation take place?"
"Do you recall the date and time of the conversation?"
"Was there anyone else present during the conversation?"
"Will you please tell the court the contents of your conversation with the defendant?"

You must be able to quote the conversation as closely as possible as to what actually took place. Remember, you must relate both your part and the defendant's part of the conversation. It is always best to refer to yourself as "I" or in the first person. The defense will not normally object if you relate the conversation in an incorrect manner, unless the content of the conversation is detrimental to the client.

It would be well for readers to realize, if they have never been examined on a witness stand by a defense attorney, that all defense attorneys are not the same. There are those who take a friendly, good-natured approach. There are those who are crafty and sly. Then there are the hell-raising, brow-beating types. The attorney's normal, real-life manner may be nothing like it is in the courtroom. Defense attorneys will act whichever way they believe is best for the defense of their client. Their job is to see that no evidence is permitted to be entered against their client which is irrelevant, incompetent, or immaterial. Defense attorneys are there to see that clients receive a fair and impartial trial. Of course, they may do a lot of objecting, but the final decisions are in the hands of the judge. Keep in mind, each registered or recorded objection, if overruled by the judge, is possible grounds for an appeal. Not only will defense attorneys try not to allow objectional testimony into the record, but they will do all they can to discredit the testimony of prosecution witnesses. They will attempt to show contradictions in the testimony of prosecution witnesses. They will also try to make witnesses testimony look faulty, weak, false, contrived, or totally unbelievable. Expect this type of behavior. Do not be surprised. Whether it meets with your approval or not is of no consequence. It is the defense attorney's job.

Defense attorneys who are doing their job will do their best to make a police officer or a security officer appear to be incompetent on the witness stand. They will employ every trick known to make officers overstate their testimony, or confuse them. They will needle witnesses in subtle ways in order to irritate or anger them. They hope such a display of lack of control will diminish the value of the officer's testimony.

Never attempt to answer a defense attorney's questions that you do not thoroughly understand. Many times an attorney will intentionally ask a vague, misleading, or multiple-meaning question in the hope you will not handle it properly and make a poor showing on the stand. Take your time before answering all questions. Give your answer some thought. There is no time limit imposed on you for your answer. If you show you are calm and deliberate and give well-thought-out answers, you will be able to take the strength out of the questions.

One final word of caution. Officers, no matter whether on the stand, off the stand, or even in the corridors, must maintain their composure and dignity at all times. They cannot relax for one moment. Jurors have the opportunity, at times, to be in the halls, on the elevators, or other places in the courthouse where they can witness the behavior of the officer. A professional, poised attitude at all times is a must for the police or security officer, especially when wearing a uniform. It is this appearance that will favorably influence a juror. Be conscious too of what you say during any recess or break in the trial. If you are in a restaurant, the hallway of the courthouse, or the elevator, you can never be sure who is nearby. It may be a juror or the defense attorney. Guard your conversation with others. During any of these breaks, you are better off not mentioning anything about the trial or the participants.

CHAPTER 15 PROBLEMS

PROBLEM 1

Security Officer Henry Champ has been a major witness in a felony trial that has been underway for the last several days.

He and another officer are key witnesses. The case is about ready for summation and being turned over to the jury, but the hour is late, so the judge continues the case until 9:00 the next morning.

Henry and the other officer remain behind at the prosecutor's request to clear up a point discussed in the trial. It only takes about five minutes, so it is only a short time before the two are in the hall, waiting for the elevator.

There are a considerable number of people in the hall waiting for the same elevator, and Henry isn't quite sure they will all get on the car or whether he should wait for another car.

When the elevator arrives, it fills up quickly, but Henry and the other officer just manage to slip into the car before the door closes.

Henry and the other officer begin talking about the matter they had just discussed with the prosecutor. The information being talked about concerns a crucial piece of information about how there had been a break in the chain of evidence by the prosecution, but how the defense had not discovered the flaw.

Both officers get off the elevator at the first floor and head in their separate directions for home.

When the trial resumes the next morning, as soon as the court is called to order, the judge declares a mistrial.

Why was a mistrial ruled in this case?

CHAPTER SIXTEEN
Firearms and Their Use By Security Personnel

The legal authority given a security officer by an authorizing agency to carry a firearm while on duty grants the officer authority to use the weapon *in the line of duty. This authority carries with it an awesome responsibility. The officer must not only be aware of this fact, but also when, where, and why the weapon may be used.*

THE HOW, WHEN, AND WHERE OF FIREARMS USE BY SECURITY OFFICERS

There can be no uncertainty on the part of security officers about the legal limits governing use of firearms while on duty. There also can be no excuse of ignorance of the times of permissible use.

You may use sufficient and reasonable deadly force only to defend yourself or another from the use of deadly force.

The use of force is an extremely dangerous area. Its importance cannot be overemphasized. All security officers must make themselves fully aware of the seriousness and possible consequences resulting from the use of force. Once again, anyone desiring to become a security officer should review the statutes of his or her own state, and that state's licensing authority. The laws of your state are what govern you.

A license to carry a firearm while on duty *does not* excuse the officer from all legal responsibilities. *Security officers are fully accountable and personally liable for any wrongful or negligent use of a firearm.*

THE RESPONSIBILITIES OF CARRYING A FIREARM

Keep in mind at all times, and under all circumstances, the license or authority to carry a firearm while on duty is not a right, a permission, an authorization, or a privilege to kill or maim someone. It is a permission granted for the use of the firearm only under the most extreme conditions. These conditions are when there is no other recourse, and only in the defense of self or others, from the use of deadly force.

A firearm is not intended to be worn as a decoration or to enhance the appearance of a uniform. It is a tool for use by an officer only as specified above. It is to remain in its holster at all other times and under all other conditions. During the course of an officer's normal tour of duty, except in the emergency circumstances mentioned, there is never a reason for the weapon to be removed from its holster.

Being accountable and responsible does not mean an officer must assume undue risks or dangers. It does mean that security officers carrying a gun must possess and be able to use good judgment. They must always be aware of the fact that *the split second one has in which to make a proper decision makes the officer the equivalent of judge, jury, and possibly the executioner.* This is an awesome responsibility and demands a deep and serious realization of the magnitude of this split-second decision. *Officers who are not able to accept this kind of responsibility should not, under any conditions, carry a firearm.*

WEARING A FIREARM IN A RESPONSIBLE FASHION

Security officers must be slow to resort to the use of firearms. They must avoid all reference to the use of firearms, under ordinary circumstances. However, they must be trained for the extreme emergency situation to be able to make quick and accurate use of weapons. It makes no sense to defend yourself or another while injuring or killing an innocent bystander. The following general rules concerning firearms should apply to all security officers.

1. Handguns carried by security officers while on duty should preferably by the revolver type, .38-special caliber, double action, with a swing-out cylinder. Heavier-caliber weapons have more ability to cause danger to others and are not necessary for defensive use. Automatic weapons have a possibility of jamming, and the need for additional rounds of ammunition is far-fetched in the realm of the security officer. The security officer's weapon is primarily for defense, not offense, as may be the case of police officers.
2. Before going on duty, officers should examine their weapon by swinging out the cylinder to verify that the weapon is empty of cartridges. Look down the barrel while the cylinder is open to see the barrel is not blocked. Load the weapon, and then close the cylinder. From that moment on, *the weapon should remain in the holster at all times, unless a real and serious emergency arises.*
3. All security officers should have their weapons checked by the agency ordinance officer or the local police authority ordinance officer upon being hired. The ordinance officer will check the condition of the weapon and either accept or reject it as worthy for service. Rejection of any weapon as being unsafe should prohibit the use of the weapon until it is made safe. When the weapon is thought to have been made safe, it should be resubmitted to the same ordinance inspector who issued the rejection.
4. All weapons should be kept clean and in perfect working order. All weapons should be unloaded and thoroughly checked to verify they are empty before any cleaning is done. No weapon should ever be cleaned or repaired while the officer is on duty.
5. All firearms should be considered loaded until thoroughly satisfied to the contrary. Accidental discharge of a firearm is considered negligence and will be dealt with accordingly.
6. When checking a revolver to see if it is loaded, *do not pull the trigger as a means of checking.* Swing out the cylinder and check both the cylinder and the barrel to see there are no rounds in any of the chambers or the barrel.
7. Never point or look over the sights of any firearm handed to you by another person. First, check the weapon to make sure it is not loaded. This check should be done as is outlined in step 6.
8. Examination of a loaded weapon should be done thoroughly and with certainty. You may get only one chance. Don't pay the price for shortcuts.
9. Never remove your weapon from its holster until you have decided it is absolutely necessary for use in an extreme emergency.
10. If you are not certain the last time the weapon was fired, check the barrel opening only after you are sure the weapon is unloaded. Be certain it is not plugged by any foreign material or slug from a misfired round. This could save your life, your sight, or some other part of your body from injury.
11. When you present a firearm for inspection, always swing out the cylinder of a revolver before handing it to another person. Let the person see the weapon is safe and not loaded. If there are cartridges in the cylinder, eject them before handing over the weapon.

THE BASICS OF SHOOTING A HANDGUN

The following basic steps may appear to be of little value in developing effectiveness in the actual use of a handgun when in a close-range situation, and under emergency conditions. Everyone knows you do not have time to concentrate on each individual instruction. These basics are mentioned so you may be aware of the proper procedures in the firing of a weapon during practice. This will help you become proficient in use of the weapon. The proper procedures will come naturally when needed in an emergency if you develop good habits with the handgun during your practice sessions.

Practice, practice, and more practice of the basics and fundamentals make possible a proficient shooter under any and all conditions. Many people erroneously believe an expert just draws and fires. This is totally false if the person expects to hit the target with regularity.

Some believe expert shooting is dependent solely upon a steady hand and a keen eye. This belief is partially true. Other factors enter into good shooting techniques, and they can only be perfected with practice. There is not a person living who can hold a handgun absolutely steady with one hand while firing it. It requires skill to fire at the split second the gun barrel is pointed perfectly at its target. There is only one way this skill can be developed: practice.

However, all the practice in the world will not produce an effective shooter if the person does not concentrate during practice. Concentration is a requirement when practicing any sport or activity in order to attain excellence, or even competence. Those who wish to shoot without bothering to concentrate during practice may as well not waste their time and their money for ammunition. Without concentration, improvement will not come.

The most common fault all beginners seem to encounter is a combination of the expectation of a loud report and severe recoil of the weapon when firing. These expectations cause what is known as "flinch." The mind imagines the report and the recoil to be many times greater than what they actually are. This anticipation makes the shooter jerk the weapon before the report or recoil actually takes place. The flinch directs the gun barrel away from its intended target before the trigger is pulled. The weapon is fired a fraction of a second after the shooter has the will to fire. In that short time the inexperienced shooter literally yanks the gun barrel away from the target during the pulling of the trigger.

Security officers should remember that they do not have to own nerves of steel or be an expert sharpshooter to outshoot the average criminal. Practice and concentration will put confidence in one's ability to meet any challenge.

Every person is different. As a result, the following fundamentals may require adjustment on the part of each individual to meet his or her need to be comfortable in what one is doing. Personal comfort is necessary for a shooter since accuracy cannot be attained while under physical strain or tension. A relaxed and comfortable shooter makes a good shooter.

The descriptions and instruction of the rudiments of firing a handgun can be practiced over and over without having to fire a single shot. The practice can take place in the privacy of your room, your basement, or outdoors. All that is needed is a fictitious target to sight on, and time. This being the case, there is little excuse for a person not to get the basics and rudiments down pat.

Position varies according to the circumstances under which a person finds himself or herself during a combat situation. In the majority of cases, where split-second draw and fire conditions prevail, it is best to face the adversary full on, face to face. The feet should be spread apart at a comfortable distance. The knees should be buckled to the point where the shooter is in a shallow squat position. After the weapon is drawn, the shooter should take a two-handed grip on the butt or handle of the revolver. Both arms should be extended to a degree that places the top of the pistol barrel on the level with the eyes so a quick sight can be made down the barrel. This is more of what is termed "point sight" than an actual aligned sight used in target or practice shooting. The upper body should be leaning slightly forward from the hips up. This makes the sighting plane and sighting an almost natural thing. This position is used mostly in close-quarter situations where time for accurate aiming is not available and point shooting takes precedence. It is also used for intermediate distances where steadiness and accuracy are required. It seems rather safe to suggest, whenever possible, a two-handed grip and squat position be used by persons in a close-quarter combat situation.

The one-hand, right-angle-to-the-firing-line position is well and good for the target shooter and certain types of competitive shooting. It plays very little part in law enforcement-type shooting, other than possibly when firing from behind a corner wall or barricade.

Take hold of the butt or stock of the revolver with the last three fingers and the back part of the palm of the hand. The tips of the fingers may touch the stock lightly so there is no strain set up in the hand, but they should not have any pressure exerted on them by squeezing the pistol grip. The thumb should rest even with or above the cylinder latch located on the left side of the frame of the pistol. Grip the pistol as high on the stock as is comfortable. The high grip will cause the barrel to point straight, almost as naturally as if the barrel were the index finger of your hand. The barrel will become like an extension of your arm.

With a low butt grip, successive shots of rapid fire will cause the gun to rise in the palm because of the recoil as firing takes place. This slippage will destroy accuracy on succeeding shots.

Use a firm grip, but do not squeeze tightly. A tight squeeze causes tension, and a tense shooter is usually not a good shooter. Try to relax the muscles in the arms and the rest of the body. If you can do this, accurate shooting will be easy.

The front sight of a revolver is usually a vertical blade positioned at the extreme front end of the barrel. The rear sight is either a notch type or an adjustable-blade type.

When sighting a handgun, the top of the front sight should be level with the top of the rear sight and an equal amount of clear space should appear on either side of the front sight blade when it is centered in the opening of the rear sight. With this alignment, the barrel of the revolver should be pointed so the aiming point, or bull's eye, appears to sit on the top of the front blade.

Careful alignment of the sights cannot be overemphasized if one is to obtain accuracy during slow fire.

It is best to sight with both eyes open if possible. Not many people are able to accomplish this because of their individual sight and eye conditions. It takes a lot of practice for those who are able to do it. This type of sighting eliminates eye strain caused from squinting or closing one eye. If the shooter is unable to shoot with both eyes open, he or she should try to close one eye softly without squeezing the eye lid tightly. This will help reduce the strain of closing one eye.

If the sighting of the handgun is done properly, the sight blades will appear sharp and in focus, while the target that sits upon the front sight blade will appear dull and fuzzy.

Under practice conditions, normal breathing should be maintained during the sight alignment. Take plenty of time to be sure of the proper sighting. When relatively certain everything is in order, stop breathing and begin to squeeze the trigger. Hold your breathing until the gun is fired. To stop breathing does not mean to inhale or take a deep breath and hold it. It simply means stop your breathing until the firing is complete.

Under emergency conditions, breathing is not usually thought about unless there is some kind of sharp shooting required. For the most part, this does not happen to a security officer. That is not to say it could not happen.

All of the prior instructions or practice will be of no avail if, at the time of firing, the trigger of the handgun is pulled instead of being squeezed. There is no one preparation for shooting that can equal or surpass the need to squeeze the trigger, rather than pull or jerk it. Without trigger squeeze any target hit is a matter of pure luck rather than ability. Trigger squeeze should become so natural to shooters that they should not be able to determine the exact instant of firing. They should be able, from practice, to determine when all of the slack is out of the trigger and be able to hold their fire should the sights leave the target.

No person is capable of holding steady the short barrel of a handgun on a small target for any length of time in order to provide consistent accuracy. The barrel will move up and down, but mostly from side to side, across the target. The sights will rest on the target only a second or two at the most. It is necessary, therefore, to develop the squeeze so when the sight leaves the target, the trigger squeeze pressure may be held until the sight returns to the target. At this time the squeeze is started again and the shot fired. Pulling or jerking the trigger will pull the shot off the target.

When squeezing the trigger, apply pressure straight back toward the body in a line parallel to the bore of the gun barrel. Increase the pressure gradually. The squeeze is to be a steady pressure application. That does not mean it must be applied slowly. It can be fast or slow as you wish or as you can master, but it must always be a squeeze and not a pull.

Trigger squeeze requires concentration and practice. This can be done with an empty weapon, away from other people. Practice of the squeeze is all-important. Unless it is mastered, hitting anything aimed at will be strictly luck.

Revolver

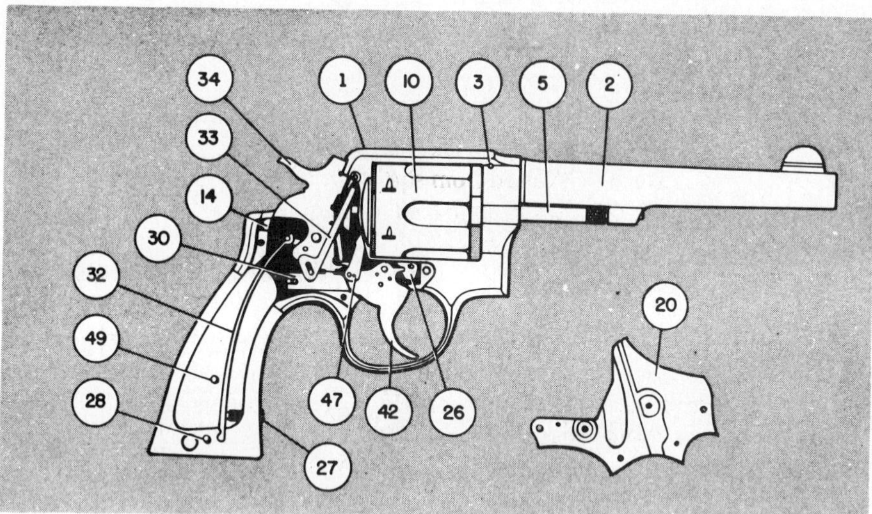

Revolver

Parts Legend

1. Frame
2. Barrel
3. Barrel pin
4. Yoke
5. Extractor rod
6. Center pin spring
7. Center pin
8. Extractor rod collar
9. Extractor spring
10. Cylinder
11. Extractor
12. Bolt
13. Bolt plunger spring
14. Bolt plunger
15. Thumbpiece
16. Thumbpiece nut
17. Locking bolt
18. Locking bolt spring
19. Locking bolt pin
20. Side-plate
21. Side-plate screws, roundhead (2)
22. Side-plate screw, large head (discontinued)

22A. Side-plate screw, flathead

23. Cylinder stop plunger
24. Cylinder stop plunger spring
25. Cylinder stop screw.
26. Cylinder stop
27. Strain screw
28. Stock pin
29. Rebound slide spring
30. Rebound slide
31. Rebound slide pin
32. Mainspring
33. Hammer block
34. Hammer
35. Hammer nose
36. Hammer nose rivet
37. Stirrup
38. Stirrup pin
39. Sear
40. Sear pin
41. Sear spring
42. Trigger
43. Trigger lever
44. Trigger lever pin
45. Hand spring torsion pins (2)
46. Hand torsion spring
47. Hand
48. Stocks
49. Stock screw

Reprinted from *Firearms Assembly 4,* by permission of the National Rifle Association.

Automatic Pistol

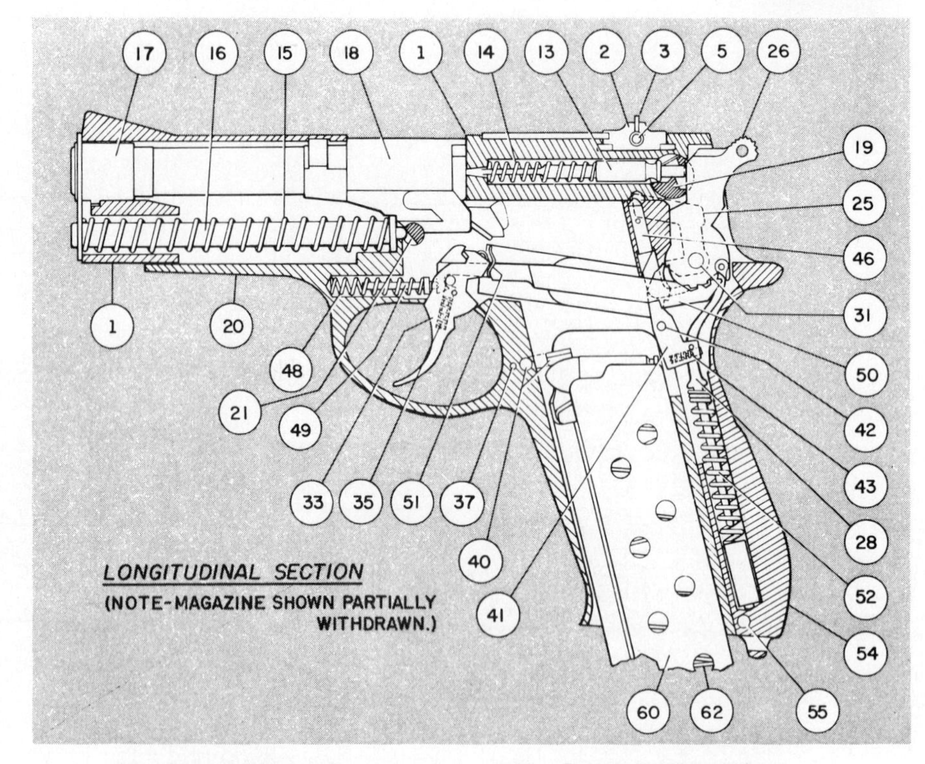

Automatic Pistol

Parts Legend

1. Slide
2. Rear sight leaf
3. Rear sight slide
4. Rear sight windage nut
5. Rear sight windage screw
6. Rear sight windage screw plunger
7. Rear sight windage screw plunger spring
8. Extractor
9. Manual safety plunger spring
10. Manual safety plunger
11. Ejector depressor plunger spring
12. Ejector depressor plunger
13. Firing pin
14. Firing pin spring
15. Recoil spring
16. Recoil spring guide assembly
17. Barrel bushing
18. Barrel
19. Manual safety
20. Frame assembly
21. Slide stop
22. Slide stop plunger pin
23. Slide stop plunger spring
24. Slide stop plunger
25. Sear release lever
26. Hammer
27. Stirrup pin
28. Stirrup
29. Ejector
30. Ejector spring
31. Sideplate assembly
32. Trigger pin
33. Trigger
34. Trigger plunger pin
35. Trigger plunger
36. Trigger plunger spring
37. Magazine catch plunger
38. Magazine catch plunger spring
39. Magazine catch nut
40. Magazine catch
41. Sear
42. Sear pin
43. Sear plunger
44. Sear plunger spring
45. Sear plunger pin
46. Disconnector
47. Disconnector pin
48. Drawbar plunger spring
49. Drawbar plunger
50. Drawbar
51. Trigger play spring (assembled to drawbar)
52. Mainspring
53. Mainspring plunger
54. Insert
55. Insert pin
56. Frame studs (4—assembled to frame)
57. Slide stop button (assembled to frame)
58. Stocks (right hand not shown)
59. Stock screws (4)
60. Magazine tube
61. Magazine follower
62. Magazine spring
63. Magazine buttplate catch
64. Magazine buttplate

CHAPTER 16 PROBLEMS

PROBLEM 1

In trying to improve the skill and accuracy of beginners in shooting a handgun held at arm's length, it is most appropriate to tell them to

A. find out which is their "master eye" and sight only with that eye.
B. recognize that their hand will shake or waver slightly, and compensate for this by trying to squeeze the trigger at that moment when their aim is most likely right on target.
C. aim at an exact pinpointed spot rather than at an approximately circular area of the target.
D. focus their vision on the gun sights, not on the target.

CHAPTER SEVENTEEN

First Aid

"First aid" is defined as "immediate and temporary care given the victim of an accident in sudden illness until the services of a physician can be obtained."

This definition covers such a wide area of knowledge that it is impossible to be all-inclusive in this manual. First aid is a major study in itself, and there will be no attempt in this text to cover every facet of the subject. This study, especially for any kind of certification, is to be found with the American Red Cross or a teaching hospital authorized to give first aid instruction. At either of these places one can obtain a basic or beginner's course, or more advanced courses that lead to advanced certification in first aid techniques.

WHAT THIS MANUAL COVERS

We will list some of the major do's and don'ts and recommend the above-mentioned courses to those who wish to become proficient in first aid practice. Training in this subject should come from qualified instructors, nurses, or doctors. Training of this type cannot be given in a manual such as this. There are some first aid fundamentals, however, that can be helpful in emergencies, and these will be covered here.

BASIC FIRST AID RULES

There are some important first aid basics that every security officer should know. Security officers do not have to stand idly by and watch someone suffer or die without trying to help. These basics are more general rules than anything else.

1. Remain calm in all situations. A flustered and excited person doesn't accomplish much and usually makes mistakes. A conscious victim seeing your own excited condition or nervousness will only increase their state of shock.
2. Control bleeding. This can be critical. Pressure bandages or finger and palm pressure should most often be used. Tourniquets should be used only as a last resort where life-threatening conditions exist and the flow of blood cannot be controlled by any other means (severed artery, amputated limb, etc.).
3. Whenever possible, keep the patient in a reclining position, making the person as comfortable as possible under the conditions present, and make every effort to calm and reassure the patient.

4. Administer artificial resuscitation and respiration, either mouth-to-mouth, mouth-to-nose, chest pressure arm lift (Silvester's method), or back pressure arm lift (Holger-Nielson method).
5. Do everything possible to prevent infection of wounds. Use compresses and never bandage directly over an open wound.
6. *Treat for shock. Shock can kill more often than the actual injuries.* Make every effort to place the head of the victim at a position below the level of the feet, and keep the body of the victim warm and free from chill. *Prevent the victim from seeing as much blood as is possible, and if the victim is conscious, give constant reassurance as to the person's well-being and that professional help is on the way.*
7. Treat burn victims for shock. Do your best to prevent infection, relieve pain when possible, and prevent air from coming in contact with the burned area. Do not apply makeshift medications or cover or touch the burn area. Flush areas caused by chemical burns with clean, room-temperature water.
8. *Do not try to treat a fracture in the field.* Leave this treatment to a doctor. Apply splints to reduce amount of movement in the limb, and treat for shock until emergency help arrives.
9. In the case of an unconscious person, call a doctor. *Do not try to administer fluids.*
10. If someone is having an epileptic seizure, *do not try to restrain the person.* Remove objects from the area where the seizure is taking place that might injure the person. Try to get subjects to bite on some object too large to swallow, but which will help keep them from swallowing their tongue during the seizure. *Do not put your fingers into a patient's mouth.* After the convulsion is over, keep subjects warm and encourage them to rest or sleep.
11. Whenever possible, use a stretcher to move an injured patient.
12. Be sure to examine subjects thoroughly for injuries before attempting any movement. If there is any possibility of internal injuries, do not move subjects unless their life is in danger from conditions such as fire or explosion. Wait for the paramedics in all but the most serious and extreme emergencies.

The following suggestions for emergency treatment and recognition of the emergency conditions indicated could be helpful to security personnel. However, *always give way to the most knowledgeable person present* when someone needs help. Be sure they get the best treatment available until professional help arrives. *Do not attempt something you don't know how to do, or don't know anything about.* You may do more harm than good. Check to see who knows how to do what.

ABSENCE OF BREATHING

If someone's breathing has stopped, artificial respiration is needed as soon as possible. A delay of more than six minutes can cause death.

Look for the rise and fall movement of the chest or abdomen to determine whether a person is breathing. Look for a change in facial color from normal to a bluish-gray color; if you place your face close to the person's mouth and nose and feel no exhaled air, you can be sure something must be done immediately. Start artificial respiration. Don't waste a lot of time looking for help or loosening clothes (unless the person is strangling). Continue respiration at the rate of 14 to 16 times per minute until normal breathing resumes. This technique should be learned from a certified teaching source, but in the absence, and as a last resort, you may consider the following methods.

The simplest and most effective method of artificial respiration is to exhale your breath into the victim's lungs. Mouth-to-mouth resuscitation can be given to a person whose breathing is extremely light or labored, but regular. Time your exhaling of breath into their mouth or nose, with the patient's inhalations.

1. If someone is available, send that person for an ambulance and for medical help, *but do not stop your efforts.* Continue them until professional assistance arrives. First, lay the victim on his or her back, and on a firm or rigid surface. Reach into the person's mouth and clear the mouth and airway of any foreign material.
2. Tilt the person's head backward by placing one hand under the head, at the base of the skull, and with that hand lift upward. Place the heel of the other hand on the person's forehead and gently push down so the chin becomes elevated.
3. With your hand on the victim's forehead, pinch the person's nostrils closed, using your thumb and forefinger. Take a deep breath, place your mouth tightly over the victim's mouth, and give four quick breaths. After this, give about 12 breaths a minute (one every five seconds). Each breath should cause a rise in the victim's chest.
4. When the victim's chest is expanded, stop blowing. Remove your mouth and turn your head toward the victim's chest so your ear is over the victim's mouth. Listen for air leaving his or her lungs, and watch the chest fall. Repeat breathing procedure.
5. Check patient's wrist or neck artery for a pulse. If no pulse is present, begin cardiac compressions, *provided you are trained in cardiopulmonary resuscitation (CPR).* This must be done in conjunction with artificial resuscitation. Continue until medical help arrives or the victim begins breathing on his or her own.

CHOKING

Obstruction of the airway by a piece of food or any foreign object is an emergency. If the airway is only partially blocked, a choking person will probably inhale enough air to be able to cough effectively. As long as a person is coughing and has good color (not bluish) you should not offer help. However, if someone is coughing only weakly, and is having difficulty breathing, first aid is needed. A person whose airway is totally blocked will be unable to speak, cough, or breathe. He or she may look bluish and clutch at the throat. After a minute or so, the person will become unconscious. If the upper airway is blocked, sweeping a finger deep inside the mouth and dislodging a piece of food may be enough to clear the obstruction. Be careful not to force the blocking particle deeper into the throat.

1. Get behind the victim and slightly to one side. Lean the person forward and support the chest with one hand. With the heel of the other hand, give four hard thumps between the shoulder blades. The victim can be standing or sitting.
2. If this is not enough, hold the victim up in a standing position. From behind, with one fist against the victim's waist area, keeping the thumb inside, hold your other hand over the fist and quickly thrust hard, in and up over the belt line.
3. This abdominal thrust should dislodge the obstruction. If it does not, repeat three times. Then, if the obstruction is still there, and the victim has lost consciousness, roll him or her toward you and repeat four back blows. If the obstruction is dislodged, give mouth-to-mouth resuscitation.

Any choking person who has been revived as a result of an abdominal thrust should see a physician. Occasionally the abdominal thrust can damage internal organs, but you should not be unwilling to use the thrust because of this.

DROWNING

In almost all cases of drowning, speed in starting artificial respiration is essential. *Do not* waste time either getting help or trying to clear the victim's lungs of water. You may need to blow quite hard, but the air you breathe into the victim's lung will pass through any water in the lungs.

If you are alone and the victim is in shallow water, start resuscitation on the spot. If helpers are available, start resuscitation while they carry the person out of the water and make him or her comfortable. Do not stop respirations even while moving the victim.

1. Start mouth-to-mouth resuscitation. Keep going until victim breathes regularly or medical help arrives.
2. When the person is breathing normally, turn the person over, raising the person's arms above the head and spreading the person's legs tightly apart. Be sure victim's head is to the side and there is nothing in the way to prevent air from getting to the person. Be sure the victim is covered and kept warm.

REVIVING A CHOKING CHILD

Sit down and place child face down across your lap. Give several thumps with heel and hand between child's shoulder blades. Be careful. These thumps should be lighter than those given an adult. An infant can be held face down with an open hand under its chest while being thumped with the other hand. A small child can also be held upside down by the ankles and *very gently* shaken. Sometimes this is all that is needed to revive a choking child.

HEART ATTACK

Heart attacks are life-threatening. Symptoms may include all or some of the following:

1. Pain in middle of chest (crushing or pressure type—severe).
2. Chest pain which moves to either arm, shoulder, neck, jaw, and even mid-back or pit of stomach.
3. Heavy sweating.
4. Nausea and vomiting.

5. Pale or discolored skin.
6. Weakness (usually extreme).
7. Bluish fingernails.
8. Shortness of breath.

Heart attack pain can sometimes be mistaken for indigestion. If there is any doubt, treat it as a heart attack. *Get medical assistance at once. Every second counts.*

Until help arrives—if patient is unconscious and not breathing:

1. Place patient on back on hard surface and clear mouth and airway in preparation for mouth-to-mouth resuscitation.
2. Follow same procedure as described under "Absence of Breathing" earlier.

Until help arrives—if patient is conscious:

1. Place patient in comfortable seated position, if possible.
2. Loosen tight clothing but keep patient warm with coat or blanket.
3. Keep patient calm and assure that help is on the way.
4. If paramedics are not available, transport patient to nearest hospital emergency room.
5. Follow same procedure if an unconscious victim recovers consciousness.

UNCONSCIOUSNESS

Coma is not the only form of unconsciousness. Unconsciousness refers to a patient in a state of drowsiness, confusion, or unable to respond to another person's presence. Head injuries or strokes could be responsible as well as excessive blood loss, lack of blood oxygen (drowning), or chemical changes in the blood.

The normal reflexes of the body disappear and muscles may lose tone and become flabby when unconsciousness occurs. The main consideration for the person confronted with this situation is an airway obstruction in the victim. This can be caused by the lower jaw and tongue flopping limply backward, thereby blocking the airway, or because the person can no longer cough and clear vomit or other matter from the throat.

Even after having treated an unconscious person, do not leave the person alone, unattended. The patient may stop breathing and the heart may stop. Should this happen, perform the same procedures as in the section titled "Absence of Breathing."

SEVERE BLEEDING

A severed or torn artery can be the cause of a rapid loss of an excessive amount of blood. Such a loss can put a victim into *severe shock* and a state of unconsciousness. The bleeding must be stopped as quickly as possible or the person may die. The loss of a pint and a half of blood by an adult, or a pint of blood by a child, is considered severe.

The goal is to stop the bleeding as soon as possible. There is absolutely no time to waste. Remember that the heart is a pump and it keeps pumping. It doesn't take long for the heart to pump out a pint and a half of blood.

1. Place victim down and raise injured part, if possible. This will slow blood flow to the wounded area.
2. Remove any visible glass or metal that may have caused the problem, but only that which is visible. *Do not probe the person's wound.*
3. Using a clean compress or pad, press down hard on wound until visible bleeding stops. If wound is wide or gaping, hold edges firmly together. If you know there is something in the wound, do not put pressure directly over wound, but around it.
4. Using a sterile compress pad directly over the wound, bind it so pressure of the compress pad on the wound is maintained. If you do not have a clean compress, there is no time to search for one. Use the cleanest possible piece of clothing (torn shirt, clean handkerchief) to make pad and binding and put into place. Infection can be treated later; death cannot.
5. Should blood seep through the dressing you have in place, do not remove it. Simply add another on top of the first one and secure it tightly.

6. If direct pressure from pads does not slow the blood flow, the next step is to leave the bandages and pads in place while applying pressure to the closest pressure point.

Applying pressure to pressure points is applying pressure to a major artery. The pressure point used should be the one located at a point between the wound and the heart. Compression of the pressure point will press the artery against a bone and should effectively stem the flow of blood to the wound area.

LOCATION OF PRESSURE POINTS ON EACH SIDE OF THE BODY

1. Under the side of the jawbone where the carotid artery passes up to the head.
2. On the underside of the arm, about an inch below the armpit.
3. Underside of the arm at the wrist, about an inch above the heel of the hand.
4. Where the femoral artery runs across the groin, before going down into the leg; on the inside of the leg where the top of the leg meets the groin.
5. Behind the leg, approximately an inch above the spot where the leg bends at the knee.

SHOCK

A person in shock is usually pale, faint, and sweating. The person has a weak but rapid pulse, and the skin is cold and moist. He or she may be thirsty or anxious, or may become drowsy, confused, and perhaps even unconscious. *Such a person needs first aid immediately.*

Shock follows severe injury, particularly burns or heavy loss of blood. Blood loss can be internal, where it cannot be seen. Thus it is critical to treat the victim so the amount of shock can be held to an absolute minimum.

1. Place victim flat, with head lower than waist and feet (legs raised approximately a foot) with face looking upward. The elevated lower part of the body will cause blood to flow from the lower body to the upper body where it is needed. Insert a cushion or clothing about six inches thick under the victim's buttox, if possible.
2. Loosen clothing and keep the person warm. Use coats or blankets, but no electric blankets or hot water bottles. The person should be kept warm, not overheated.
3. Do not administer food or water.
4. Get professional help as soon as possible.

CHAPTER 17 PROBLEMS

PROBLEM 1

There is a minor fire at St. John's Hospital, and Security Officer Henry Lamar is struck in the eye by a flying object caused by an explosion. Lamar's eyelid is torn, but his eyeball is not injured.

The best thing to do in such an emergency until professional help is available is

A. Swab the injured lid with boric acid solution.
B. Swab the injured lid with alcohol.
C. Wash out the injured eye with very dilute phenol solution.
D. Apply a loose sterile bandage.
E. Apply a tight sterile bandage.

PROBLEM 2

In a fracture, all but one of the following procedures are essentials of first aid:

A. Do not move victim. This could encourage further injury.
B. Treat for shock.
C. Control hemorrhage, if present.

D. Immobilize with splints.
E. Reduce swelling by bathing with warm water over the painful area.

PROBLEM 3

What special infection dangers are puncture wounds subject to?

A. There is generally a fracture near the wound.
B. The patient suffers from shock.
C. They usually don't bleed profusely; thus cleansing caused by bleeding is not present.
D. It is hard to stop the flow of blood from a puncture wound.

PROBLEM 4

When treating a first aid victim who may be in a serious state of shock, which of the following is generally poor procedure (unless done to prevent further injury)?

A. Keep the victim warm.
B. Keep the victim quiet.
C. Give fluids in small amounts.
D. Elevate head and shoulders.
E. Stop bleeding.

CHAPTER EIGHTEEN

Fire Prevention

Fire prevention is a major function of the security officer while on duty. The possibility of fire should be a matter of top priority in the mind of officers during their tour of duty.

If the security officer is assigned a post where the client has an emergency plan in effect, drawn up by the administration of the facility, it becomes the duty and the responsibility of the security officer's employer to determine this fact when taking the contract, and obtaining a copy of the plan. The contractor should extract the part the security force is to play in the master plan and see that each officer at the facility is given a written copy of his or her duties in such an event. It is wise to place a written copy of the duties at each individual security post in the facility.

The master plan must be followed by everyone if it is to be successful. Security officers cannot be running around on their own, doing what they believe to be helpful, but actually acting contrary to the master plan.

FIRE PREVENTION

Fire prevention is a different matter, however, from actions to be taken during a fire. If the security officer is doing the proper job he or she is assigned to do, there may be no need to activate a master emergency plan. It is possible, by both their actions and their alertness, that officers can prevent a catastrophe before it ever has a chance to start.

The following information is for the security guard doing duty where there is no master emergency plan in effect, or in a place where there may be only a small number of persons working or living, or perhaps a place devoid of any people.

Fire prevention is defined as the reduction to a minimum of all possible causes of fire. These unwanted fires are usually, but not always, the result of carelessness, ignorance, or malice. Realizing this to be the case, it behooves security officers to make some immediate checks of the area around their post as soon as they report for duty. *These checks should be made each and every time they arrive for duty, no matter how long they have worked the post. Things can change drastically within 24 hours.* During the guard's absence while off duty, some carelessness or malice may have taken place. The checks the officer makes are for the client's welfare and also for the officer. He or she may be the first person affected by any incident that occurs while on duty.

A security officer's first duty, in the event of fire, is to turn in the alarm. Next, if possible, is to begin the evacuation of all personnel. After being certain all personnel are removed from the area, administer first aid to any who may need it as a result of injury or smoke inhalation. Should evacuation and first aid not be required, the officer should then employ any fire-fighting equipment handy and attempt to contain the fire until professional assistance arrives.

Once a fire begins, it is too late to start searching for the things one should have sought out before beginning the tour of duty. It is too late to start looking for a phone to call 911 for assistance. It is too late to see how many persons are in the building who need to be evacuated, and the areas in which they are working or located. It is too late to find out where the fire-fighting equipment is located. And it is too late to locate exit doors that may or may not be locked, or to determine where the doors lead to.

Alert security officers make sure they know all of this information as soon as they report for duty. They make it their business to ask questions about the things they don't know, such as whether the client uses volatile materials, explosives, or substances that give off hazardous fumes should containers leak. Officers make it their business to check the personnel working in the immediate area of their post, where the exit doors are located, where the fire-fighting equipment is located, and locations of the phones. Once they are certain of all of this each time they report for duty, they make a visual check of the equipment. Officers check the doors for blockage, the area for improper storage, and whether the phones are indeed working.

Fire causes loss to a client, even if there is insurance coverage, because the interruption of business caused by the fires is seldom compensated for in full. Some clients carry business interruption insurance, but the chance of their doing so is not very high. The cost of this type of insurance is expensive. Even having business interruption insurance cannot compensate for the inconvenience of the loss of important business and accounting records.

To do all they can to reduce the chance of loss by the client, it is necessary for security officers to have a working knowledge of the ways fires can start, and the type of fire they are dealing with. Each type of extinguishing material and the use of the wrong material on a particular type fire can be extremely dangerous to everyone involved.

Security officers do not have to be fully trained firefighters. If this were the case the officer would most likely be a member of the fire department, whose members earn more money. It is necessary, however, for security officers to be able to act as "stoppers," or to be able to take stopgap measures until the professionals arrive.

If during a particular round on a post or area the officer detects what is a serious fire or safety hazard, he or she should immediately make a written report of the situation to the client. Officers should give copies to their employer and retain a copy for themselves. If something isn't done about the report and action taken to correct the situation, they should follow with a second report. If the situation is really bad and the client does nothing about it, the officer should ask the employer to speak to the client about the matter. Keep copies of the reports and any actions taken. If the fire marshal should make an inspection and find the same problem and the officer did not report it, the officer and the employer could be in serious trouble.

THE FIRE FORMULA

There is a formula for a fire just as there are formulae for various mathematical problems. In this case, the formula for fire is a simple equation.

Heat + Material + Oxygen = Fire

Remove any one of the parts of the above equation and there is no fire. Take away any one of the ingredients and the equation is reduced to where it is not able to produce the end result.

1. Heat removal (or cooling)

   On class "A" fires (wood, rubbish, textiles, etc.), saturating with water lowers the temperature of the material below the burning point. For class "A" fires (described below) *use water or foam.*
2. Material removal (or starving)

   Elimination of the supply source of the fire, such as gas, oil, lumber, etc., from the area of the material involved, provided it can be done safely, will starve the fire. Removing all excess burning substances causes the fire to starve.
3. Oxygen removal (smothering)

   Depending on the type of fire involved, the use of chemicals, fog, sand, water, black powder, or other materials can reduce the oxygen content below normal, and the fire will extinguish itself. Without a supply of oxygen, a fire cannot burn.

Fires are classified by type, and for each type of fire there should be a particular type of retarding agent used. Knowledge of the use of the proper agent can save your life and the lives of others. To use the wrong retarding agent on a fire can lead to disaster.

EXTINGUISHERS

Before using a fire extinguisher, be certain you know the type of extinguisher you have, the type fire it should be used on, and the type of fire you are confronted with. Do not grab an extinguisher, thinking it doesn't make a difference, and start spraying the fire. It could be a mistake.

CLASSIFICATION OF FIRES AND TYPES OF EXTINGUISHERS

Class "A" fires — These fires involve wood, rubbish, textiles, straw, weeds, and so forth. Use water or foam.—Do not use CO_2, carbon tetrachloride, or black powder.

Class "B" fires — These fires involve oils, solvents, and grease. Use CO_2, foam, carbon tetrachloride.—Do not use water or black powder.

Class "C" fires — These fires involve electric devices—house wiring, electric motors, automobile wiring, etc. Use CO_2 or carbon tetrachloride.—Do not use foam, water, or black powder.

Special class — This is a special condition, not contained in any of the above, but within the standard fire code. It is fire involving magnesium or titanium alloys. This material is usually found in or around machine shops or metal foundries. It is usually ignited by heat or friction caused during a machining or forging operation, and is extremely volatile. Pure raw magnesium or titanium, without the benefit of being alloyed with another metal, will ignite in the air. It is very dangerous. Magnesium and titanium alloys are most often used in the aircraft industry or where the metals used must be light but yet have strength.

During the burning process, extreme caution must be used by anyone attempting to extinguish these fires since small pieces of white hot material are constantly exploding, sending splatters in all directions. These particles can cause deep, penetrating, and very painful burns. The heat from this type of fire is greatly in excess of the heat caused by most other fires. *Caution:* Use black powder (or if not available, sand or dirt). This type of fire *must be smothered.* Do not use foam, water, or carbon tetrachloride. Any of these will make the explosions more severe, and possibly put the fire out of control. This is an extremely dangerous fire.

Use only CO_2 on electrical fires. CO_2 is not a conductor of electricity. Your life can depend on using the correct extinguishing agent in an electrical fire. Carbon tetrachloride or pyrene is permissible, but are usually not available. They have almost exclusively been replaced by CO_2. *Beware:* If you do use carbon tetrachloride be very careful; it can emit dangerous fumes.

APPLICATION OF EXTINGUISHING AGENTS

When applying black powder to a titanium or magnesium fire, be sure and cover the entire top of the mass, along with all of the outside edges. Application of the material to the burning mass should be done with extreme caution since both titanium and magnesium burn with an intense heat, and both have a tendency to spatter particles all over the area. Burns caused by these spatters are very serious because they burn deep and are difficult to heal. Any moisture or water applied to either of these burning materials makes the spattering worse. An explosion from the application of water could easily put the fire out of control. The whole idea of the application of the powder, sand, or dirt (or of all three) is to smother completely the burning material. *Do not disturb the covering, once applied,* to see whether the material underneath is still burning. It takes a considerable time for the material to be extinguished, and opening it to air will just start things all over again.

When applying the extinguishing agent to an oil fire, do it gently from the edges in toward the center of the fire. Pressure from an extinguisher can cause the fire to spread. The agent should be applied as directly as possible, and yet in sufficient quantity to cover the burning material completely. The softer the force of application of the extinguishing agent the better. It requires very little force or pressure to cause oil to spread.

Most computer facilities have halogen-type extinguishers available, or the halogen is piped directly into the system. This special material is used in computer fires to minimize the amount of damage to the expensive equipment and records. It has an evaporation quality that permits the extinguishing of the fire while leaving very little residue.

THINGS TO DO IMMEDIATELY UPON ASSUMING YOUR POST

1. Locate the nearest fire or alarm box (if still in use).
2. Know the location of the nearest phone, and the fire department number; or use 911 if available in the area.
3. Check out corridors to determine where they lead.
4. Check all exits. See that they are not blocked or locked.
5. Locate the nearest extinguishers or fire apparatus, and determine their type.
6. Determine whether other persons are working in the building or are there for legitimate purposes.
7. Check all fire doors. Make sure they can be opened from the inside and are not chained or locked in any fashion, inside or out.
8. If the post is in a factory or warehouse, determine the type of products manufactured or stored there. See whether there are any drums or containers of volatile or explosive materials.

SECURITY PRECAUTIONS

The best time to stop a fire is before it starts. The officer should look around the general area, taking particular note of the general housekeeping conditions. Poor housekeeping can be a major contributing factor to fires. They may not start the fire, but they sure help it to spread.

A written report of poor housekeeping conditions should be made by the officer to his or her superior, and a copy of this report should be forwarded by the officer's employer to the proper management authorities of the client.

Enforce "no smoking" regulations if the client has them in effect. Make no exceptions.

Check to see that flammable liquids are kept in closed containers, and in a well-ventilated area. Fumes are sometimes a more powerful explosive than is the material emanating the fumes. It would not hurt for the officer to submit a suggestion to the management to provide exhaust fans for this area. They can do no more than say no.

Keep in mind, storage areas that are enclosed can sometimes overheat and spontaneous combustion can result. This type of ignition seldom, if ever, gives any warning. It must be detected by constant vigilance.

In areas containing grease and oil products, make sure that oily rags are stored in covered containers, not left out in the open on machines or on the floors.

A very serious condition often overlooked is dust of any kind. Some dusts are more dangerous than others. The client's products may give off dusts that could be highly explosive. The right amount of a particular dust, mixed with the right amount of oxygen, can be the cause of serious explosion and even fire. Dust explosions are formidable. Graphite dust is possibly one of the most dangerous types. These explosions give no warnings and are truly devastating. Grain-mill explosions are an example of a dust explosion. In areas where serious quantities of dust particles abound, make certain there is plenty of ventilation.

CHAPTER 18 PROBLEMS

PROBLEM 1

Security Officer Williams is making a tour of the shop area of the Smith Aircraft Company when he comes upon a white-hot fire burning in the bed of an engine lathe. As she approaches the area she notices the fire is burning with a brilliant white glow, is radiating terrific heat, and is sputtering small particles of burning, molten material around the immediate area.

What type of fire is indicated and what type of extinguishing agent material should be used?

CHAPTER NINETEEN

Traffic

This book does not recommend or suggest that security officers must be able to direct traffic or that they be traffic officers. What it does suggest is that the time may come when some knowledge of traffic, on private property or otherwise, may be necessary and helpful. This is when the security officer should have at least the basics of traffic control.

Security officers should have sufficient knowledge of the fundamentals to help out in an emergency if called on by the police for aid, or to handle large parking lot assignments on private property. It is important to be able to handle an ordinary flow of traffic. There is nothing like angry motorists caught in a traffic jam. They lose their personality and often become vicious.

Traffic direction takes practice to become proficient. But proficiency in traffic direction is not necessary. Knowing the basics and being able to help out where needed is sufficient. Take advantage of any opportunity to practice in real traffic situations.

Certain things should be kept in mind when you take up a position in an intersection or any other traffic location. You have unwillingly become what might be termed "fair game" for all of the "poachers" who are driving around in the "asphalt jungle."

Be very certain you are able to see, and to be seen, by all approaching lanes of traffic, and by all pedestrians. If you are working after dark, make sure you have a flashlight and are wearing some items of reflective coating. Make certain, from your position, you can assure free movement of traffic and are not be more of a hindrance than a help. Remember, from the spot on which you stand you must be able to control the flow of traffic, permit turns to be made, and assure the welfare and safety of pedestrians. That spot on which you stand must be in a location that provides safety for you, and not let you wind up as a hood ornament on some poacher's car.

BASICS OF TRAFFIC DIRECTION

The following basics are sufficient to keep you out of trouble and to keep traffic moving in a reasonable manner.

1. Use clear, definite, large hand signals and gestures. Be certain the movements are uniform and indicate to drivers, beyond a doubt, what it is you wish them to do.
2. Always try to make traffic interruptions for traffic flow changes during a natural lull in the flow. By doing this, you reduce the chance of rear-end collisions.

3. If the traffic is so heavy not to allow natural gaps at periodic intervals, use a loud blast on a whistle, and long and slow arm movements. In a pronounced and clear manner, point directly to the vehicle you wish to stop. Do not wait until the vehicle is almost upon you to try to make it come to a halt. Give the driver plenty of warning, keeping in mind you may be in danger, and so too the rear end of the oncoming driver's vehicle.
4. Should a large truck be in the line of traffic, try to let the truck through before you stop traffic. In this way, when the traffic starts up again, there is no delay because of a slow-starting vehicle such as a truck.
5. Keep the traffic moving and the drivers alert by the use of short "toots" on your whistle, and by clear, obvious signals.
6. Should a tie-up develop, don't get excited and panic. Keep your head and your temper. Look over the situation, pick out the trouble spot, make up your mind how you will clear the situation, and put your plan into action.
7. Above all, don't leave your position with the intention of bawling out some driver. In most cases you will not have the authority to do so.

The basics outlined above should permit you to regulate the cross flow of traffic effectively, reroute or detour traffic, assist pedestrians, and prevent blockage of traffic.

USE OF HAND SIGNALS

Clear, obvious, unambiguous hand signals are all-important to traffic control. These are the means by which you convey your instruction to the drivers of vehicles. If your signals are not clear and concise, you could possibly pay for your sloppiness with your life. *Drivers are not able to read your mind or hear your voice, so take heed.*

To halt traffic, point your arm and finger directly at the vehicle you wish to stop. Keep your finger pointed and your eye on the driver of the vehicle. Give a long blast on your whistle to be sure you have the driver's attention. At the same time you are blowing your whistle, begin to raise the fingers on your pointing hand to a vertical position, so the palm of your pointing hand is facing the driver. This signals the driver that you wish him or her to stop. Hold your hand in this position until the vehicle comes to a halt.

When desiring to stop the flow of traffic coming from the opposite direction, in preparing to allow cross-traffic flow, turn your head to the new direction while holding the stopped traffic with a continuation of your original movement. With your free arm and hand, repeat the same procedure with the new direction of traffic. Do not lower your arms and hands until you are certain the traffic you intended to stop, in both directions, is definitely stopped.

To start cross traffic in the opposite direction, turn your body so your side is facing the new direction. Hold one direction of the already stopped traffic with a raised hand stop signal, and with the free arm and hand, indicate with your pointed arm and finger that you wish the new direction of traffic to prepare to move. When you are sure you have the driver's attention, with the palm of your hand in the upward position, raise the forearm and hand in an upward direction, bending at the elbow, and passing the arm in front of your face. When the traffic in the direction starts to move, repeat the movements for the traffic in the opposite direction.

CHAPTER 19 PROBLEMS

PROBLEM 1

Why can the flow of traffic, at a busy intersection, be better controlled by an officer than by a traffic signal that is set to change at predetermined intervals?

A. Some people are colorblind.
B. An officer can start and stop traffic as necessity demands.
C. An officer can stop suspicious vehicles.
D. An officer is easier to see than a traffic light.

CHAPTER TWENTY
Patrol

Protection of life and property is the standard role the security officer plays in everyday life in the United States. Patrol objectives are in the security service what "general practice" objectives are to the medical profession, namely "all-inclusive."

PREVENTIVE MAINTENANCE

Probably one of the most important security objectives is what is known as "preventive maintenance." There must be a continual dialogue with many individuals from many disciplines to discuss and explain the role of security. There must be an exchange of views on the merits of respect for the law and the rights of property owners and others. The discussion must include the manner in which the laws are enforced and rights respected. This dialogue is security work, and it is an extremely important part of it. Too often this function is labeled as "public relations" and is shoved into a corner to gather dust.

Preventive maintenance in the security business requires an approach wherein the individual security persons interact with the public to develop and maintain a rapport, and to earn respect for their efforts.

MISSED OPPORTUNITIES

Security officers on patrol make personal contacts with the people they encounter on their beat on a one-to-one basis. They should take advantage of every opportunity to permit these people to know them and their work policies.

Unfortunately, too many people have developed opinions and attitudes about security personnel solely on the basis of a chance contact with guards who are indifferent to the public. The resultant opinion of such a contact is that all security personnel feel the same way and have no interest in the welfare and property of those they are assigned to protect. Patrol, when put into operation, provides an opportunity to overcome such unfavorable attitudes. There is no better way to cultivate good feelings than through personal and informed contact with people, and it is during these contacts that officers can exhibit an attitude and an appearance that win respect.

CRIME PREVENTION

"Crime prevention" is the phase of a security officer's role in law enforcement in which the officer attempts in as many ways as possible to eliminate or reduce the desire of others to commit crimes. In the security officer's part in the system, major crimes against people seldom come into play. That is not to say it could not happen. The security officer deals mostly with crimes against property and misdemeanor offenses.

Security officers on patrol, be it foot or mobile, are distinctively marked by a uniform, and if in an automobile, a marked patrol car. They are immediately associated with someone responsible for the enforcement of the law, the rights of private citizens, and the safety of businesses within the confines of their patrol area. Their high visibility is a deterrent to crime.

THE PATROL'S CONTRIBUTION TO CRIME REDUCTION

A security patrol's contribution to reduction in crime occurs when the patrols reduce the opportunity or change the mind of others to commit a crime. In earlier times the night watchman of the neighborhood walked the streets and alleys at night with a steel-tipped nightstick. His tapping noise, as his stick struck the pavement, warned those with ideas of doing something wrong that he was in the area and ready to act. The "tapping watch" was a method of keeping crime down during those times. Today this would not do. Highly visible marked cars with the ability to cover ground faster, highly visible uniforms, radio communications, and irregular patrol patterns are the methods used. This system creates a feeling that officers are more apt to be in the place they are needed, when they are needed.

The theory of patrol is based on the hope that individuals contemplating committing a crime will not go through with it knowing someone with authority to legally stop them is nearby to do so. The uncertainty of where the officer is and where the officer will appear reduces the individual's desire even more. Perhaps the presence of threat or presence of one possessing the authority to interfere may not keep the subject from planning a crime, but it can minimize the opportunity for carrying it out.

THE FUNCTION OF A PATROL

Most businesspeople take pains to secure their place of business from theft and vandalism. They apply special doors and locks, but sometimes they make errors in judgment in the interest of economy. Some neglect to give the same precautions to rear doors, which customers never use. These doors may have hinges on the outside, thus allowing the removal of the hinge pins without any trouble and entry by intruders. They put in alarms, cover skylights and ventilation shafts, and install other alarm devices. With all of these precautions, however, it is not unusual for someone to leave at night or over the weekend and forget to set the alarm or lock a door. Professional people such as doctors and lawyers often work odd hours, and it is not unusual for them to leave their offices unlocked and vulnerable to burglars. To compensate for these lapses, patrol officers check buildings and other property during their beat on an unscheduled and frequent basis. They check the doors and often apprehend the person who tries to take advantage of such lapses in security.

A patrol operation has the responsibility of inspecting for security of the client's premises, but there are also other services that are performed. Officers inspect for fires, fire hazards, the effectiveness of provisions made for security purposes, for code violations, and for safety hazards. They report their findings to their immediate supervisor along with furnishing a report to the client.

The small things a patrol officer looks for are broken water lines, downed telephone poles and electrical lines, conditions of streets in their patrol area, health and safety hazards, or violations of any ordinances or building codes, safety codes, or fire codes. Reports of any of these conditions should be made immediately to one's supervisor, with copies once again going to the client. All of these observations and inspections are performed by an alert security patrol performing their best for the client.

Patrol officers have a duty to establish personal contacts within the patrol route. They cannot be satisfied with their own knowledge that they are doing a good job since this opinion is biased. There are others who observe officers in their daily routines and know how and why they perform their job the way they do. They should be able to evaluate their performance as well or better than the officers can themselves. A good rapport and a meaningful interaction between security officers and those they serve is a necessity. Officers must have the support not only of those they are employed to serve but also as many others as possible.

FOOT PATROL

Foot patrol is the most basic type of patrol. Normally it is used in large office buildings, warehouses, shopping malls, and even small parking lots or garages. Although it confines an officer to a limited area of coverage, the foot patrol is still among the most effective of the various types of patrol. Foot patrol consists of different methods: the fixed post, the line beat, or the random patrol. Most of the time the foot patrol is intended to deal with special problems or conditions particular to an area or location. The moving patrol is used primarily in areas where there are considerable crowds in confined spaces such as shopping malls and sometimes large department stores. Fixed or stationary patrols are normally the assignment for the entrances to office buildings, individual offices, loading and shipping docks, and entrances and exits. Random foot patrols are mostly used in factory or warehouse complexes, the walkways of shopping malls, large parking facilities, or apartment complexes. For the most part, it is best not to have specified rounds in the moving foot patrol unless there are clocks to punch, or specific fire or safety items that must be checked on a routine basis. The random foot patrol should be on the constant move, with no routine or time schedule. No one should be able to set his or her watch by a moving patrol.

HORSE OR SCOOTER PATROLS

When very large areas are involved, it is sometimes advisable to consider horse or scooter patrols. They are especially good in large apartment complexes, estates, extensive park areas, and large exterior mall parking lots. This type of patrol is very successful in the prevention of vandalism, trespassing, and in protection against assaults. Horses are especially good when there are no hardened surfaces for the motorized vehicle to run on. The motorized scooter is especially good in foul weather since they are, or can be, enclosed and even radio equipped. They provide more space for an officer to carry report forms, foul weather gear, and other equipment.

AUTOMOBILE PATROLS

Automobile patrols are used when there are very large areas to cover such as industrial complexes. It is the officer's job to check for signs of forced entry, fire watch, answering burglar alarms, destruction of property, and suspicious vehicles such as trucks attempting to move or remove unauthorized materials. Automobile patrols are effective when there is more than a single client and when much area is covered.

In the automobile patrol, there is a possibility of conflict with the local police if the patrol officer or officers are not properly trained, if they exceed the limits of their responsibility and authority, and if they fail to call in the police when it is proper to do so. The contract company furnishing the auto patrol should coordinate with the local police and familiarize them with their client, who the officers are, how they have been trained, and how they have been instructed to cooperate with the police. It is true that on private property the security officer usually has the same authority the police have, but after the initial occurrence of any incident, it must be remembered the incident becomes a police matter. To have poor communications or lack of cooperation with the local police is like shooting one's own foot.

THE MOBILE PATROL OFFICER

The mobile patrol security officer is not a police officer, and is never to think otherwise. Security officers must know the limits of their authority and responsibility and never intentionally exceed them.

RESPONSIBILITY AND AUTHORITY

Security officers have the responsibility and the authority within the scope of their license to act and react in a variety of ways when encountering violations of the law or infractions of the client's rules within the confines of the premises the officers are licensed and contracted to patrol. Officers may contact suspected violators and admonish them without arrest, or they may arrest violators if circumstances require an arrest. In either of these situations, officers must write a report of the circumstances involved, the reasons they acted the way they did, and a detailed account of all actions taken. In the event of an arrest, the person or violator is to be detained for the local jurisdictional authority, who will take over. At the point of taking custody, the police have the responsibility for the care of the person in custody. In the case of an arrest, a report should be written with the idea in mind that the report will become a supplement to the police officer's report. Care should be taken in the form and content of the report. If the situation is one in which a contact was made and no arrest or detention was effected, the report should be made in detail and turned in to the supervisor and the company for further action and for their records.

PRIOR TO STARTING PATROL

Prior to starting a patrol, security officers should arm themselves with the knowledge of what took place in their assigned area since they were last on duty there. If trouble spots have developed or if a client has requested a special service or issued particular instructions that are in addition to, or different from, normal, the officer must know about them. This information should come verbally from the officer being relieved, and from written instructions contained in a log book or a special instruction form issued by the employer.

During the vacation season and during holidays, the officer should be made aware of special closings of companies, or if patrolling a residential area, the people who are on vacation, the time they left, where they can be reached in an emergency, and when they will return. This information should be contained in a log book kept in the patrol car, or on special company forms. Special window, door, and premises checks should be made a number of times during each shift or tour of duty. These checks should be made physically, not just visually. If anything is found unusual, the facts should be recorded in the log book along with the name of the officer who found it, and the date and time it was determined. If the matter is serious enough, the company management should notify the property owners.

COMMUNICATIONS WITH LOCAL POLICE

Periodic checks with the local police in the area of the patrol should be made. The police have daily contact with the crime reports covering not only the patrol area but all areas surrounding it. They usually know, in detail, any influx of a particular type criminal into the vicinity. They will keep the patrol advised of any changes. When so advised, officers should note this information in the log book for the benefit of those who follow them.

CHECKING THE PATROL CAR

The patrol officer, prior to assuming command of the patrol vehicle, should inspect it for general conditions, cleanliness, and workability. Any faults, no matter how insignificant, should be recorded and passed on to the supervisor. If a condition is noted and recorded before you use the vehicle you cannot be held accountable. If you do not record discrepancies, and the person who relieves you does report something amiss, you may have a hard time explaining why you are not responsible.

THE FIRST TOUR OF A SHIFT

The first tour you make on your shift should be made with much deliberation and attention. If there is anything out of place or amiss, this is the time you want to discover it. You don't want to be halfway through your shift before you become aware of something wrong. If you find a discrepancy on your first round, you will know it took place sometime before you arrived or since you began your patrol. The time element in an event could be of great importance to both the police and the client.

A log should be kept for each shift of a patrol. This log should contain all occurrences during a shift, be they large or small, or in any way different from the normal. It should contain the name of the officer reporting for duty, time in, and time out, along with any special instructions. The officer should make an entry to the extent "everything secure" if there is nothing else to report about the shift.

STAYING AWAKE

In a mobile patrol there are times when driving becomes very tiring. During the nighttime hours drowsiness can be brought on by the monotony caused by lack of activity. An officer can be lulled into complacency and sleepiness quite easily. This is all the more reason to get out of the vehicle to make the physical building checks. You will not only do a better job but you will stay alert. If you find yourself getting sleepy, open the window of the vehicle and let the fresh air revive you. If you're alone, talk out loud to yourself occasionally. Stop ever so often, get out of the car, and stretch your body for a few moments. Never use medication or pills to stay awake. Adequate rest during your time off is better than any chemical crutch. If you are not able to handle a certain shift in a manner equal to what is required of an alert security person, request a change of shift. Do not run the risk of letting yourself or someone else get hurt because of your inability to stay awake and alert.

BUILDING INSPECTIONS

During a tour of duty, the patrol officer performs inspections for a variety of purposes. Fundamentally, inspections of buildings and other places involve building security and early discovery of fires and other hazards. Additionally, such inspections serve as a crime deterrent or prevention tool. There is always an extra deterrence factor during the actual presence of the officer.

While on patrol, get out of the unit and physically check buildings. From the front seat of a patrol car, with the aid of a powerful spotlight, it is possible to check windows and the ground below for broken glass, but you can't see tool marks or evidence of attempted entry. Being afoot and placing a beam of light on the space between double doors, it is possible to see the locking bolt in place or out of place. One technique used by officers for building security is to get out of the car and try the doors and windows of the building, then place a small piece of paper in the crack above the door or window. When returning periodically for another check, the officer shines the light on the slip of paper from his or her place in the car, assuming the building is secure since the paper doesn't appear to be disturbed. Many enterprising burglars have known this trick for some time. They note the position of the paper, pry open the door or window, burglarize the premises, and then put the paper back in place. The paper never leaves, but the security has been violated and the patrol officer doesn't realize it. This system of security is not recommended.

Plan to check out the entire group of buildings when you approach a shopping center or complex of buildings. Although you may use a spot-check technique, you should be sure of the security of the entire area. How often or how much you check such an area depends on the contract your employer has with the client. If the client wants an extremely large area or complex covered with only one patrol, spot-checking or the checking of specific areas may be all that is possible. If the patrol covers the properties of a number of clients within the same patrol, it is necessary and required that each client's property receive the same treatment. Individual checks of individual properties are necessary. This imposes the obligation of your employer to limit the coverage required of any one unit. To include an unreasonable number of clients in one patrol does an injustice to the clients and to the security officers responsible for the proper protection of the property.

PROPER CHECK-OUT PROCEDURE FOR BREAK-IN OR BURGLARY IN PROGRESS

If, during your check of a building, you have reason to believe there has been a security violation of the building, and there is a possibility of someone being inside, do not immediately enter the building to check. First, notify your dispatcher of what you have found, your location, and ask for the local police to be notified so they can send back-up assistance. Then return to the scene and take up a position near what you believe to be the point of entry. Await the arrival of the police. Should a suspect attempt to escape through the original point of entry, you will be in a position to apprehend the suspect and hold him (or her) for the police. Upon arrival of the police, signal them of your location by use of a flashlight or by making yourself visible. Explain to them what you have found and the reasons why you believe the building is still occupied. Take further instructions from the police.

Should it be you are unable to communicate with the dispatcher or the police and find you must go it alone, use extreme caution and proceed as follows:

First, park your unit some distance from the building to be inspected and approach the rest of the way on foot. If someone is inside, there is less likelihood he will hear you approach on foot. Move silently, minus the rattling of keys or the tap of leather heels. Walk quietly, placing your entire foot on the surface of the ground at the same time to assure less noise. Take advantage of the natural cover of darkness and shadows. Do not use the flashlight except when absolutely necessary, then only when holding it to one side, away from your body. Walk close to the buildings and avoid silhouetting yourself in doorways or in front of windows.

Plan your approach so you are not walking in the glare of the headlights or other light, or casting shadows to signal your presence. If you are fortunate enough to be working with a partner in a two-person patrol, you and your partner should plan your strategy to complement each other. Both of you should know the position of the other at all times. One effective way of communicating with other officers is by tapping short code signals on the pavement with a baton, flashlight butt, or other hard instrument. Other methods include flashing a light on a high object, which may be visible to officers on two sides of the building, or flashing a signal directly toward another officer with the hand covering all but a small "pinhole" of light coming from the flashlight. If you must talk, whisper. Whatever system you plan to use, be sure your partner is aware of your intention and is clear as to the manner of its use.

If there is reason to believe a building is harboring a burglar or other felon, approach the premises with caution, being alert to everything around you and to every sound. Though some may not agree, to approach the building with a drawn weapon at this time is a questionable technique and should be avoided. The best place for a weapon at this time is in its holster. You already have one hand holding a flashlight. The other should be ready for doorknobs, window latches, and other objects you may have to handle while checking the premises. It takes very little time, if any more, to draw and fire a weapon than it does to fire it when it is being held in the hand. If an armed suspect is inside and intends to shoot you when you enter, and if you have absolutely no knowledge the suspect is inside, there is no reason to believe you would prevent the suspect from shooting at you by having a gun in your hand. The intruder most likely knows where you are while you do not know where he is. He has the drop on you in this instance and your holding a gun will not help you in any way. Consider, also, that if you are turning a doorknob with a gun in the same hand and a burglar on the other side of the door suddenly kicks the door toward you, you are in a vulnerable spot. A free hand, up until the time you actually confront your opponent, can be a great advantage.

While approaching the premises, look around for vehicles that seem to be out of place, or sacks, boxes, or suitcases that may have contained burglar tools for roof or safe jobs. Check for footprints in any mud or in loosely packed earth beneath doors and windows. These may indicate a point of entry or exit. Inspect the sides of the building where it is adjacent to a utility pole, piles of boxes, ladders, or other means of access to the roof.

Make a serious task out of your building inspections. Vary your technique and starting point each time. You will generally cause confusion for anyone who might attempt to analyze your methods. The rear doors and windows are most vulnerable to attack by burglars. Some burglars prefer the front because they believe officers only check back doors, or doors that are not visible from the street. *Never overlook the possibility there may be some person inside who has a legal and legitimate reason for being there. However, consider everyone suspect until you have completely checked them out.* Also consider the possibility a lookout accomplice may be sitting in a parked vehicle somewhere nearby. The accomplice may have all of the outward appearances of someone who has a perfectly logical reason for being at the location. The lookout may have some unique method for signaling your presence to the burglar inside, which may indicate to you his possible complicity in some kind of criminal activity.

When checking out a door, look first to determine which way it swings open. Some burglars close the door just far enough so when someone carelessly grabs it to see if it is locked, the bolt locks in place. The clicking sound of the bolt serves as a signal device for a burglar. *Do not stand directly in front of the door;* the intruder may shoot through the door. Hold the door steady, grip the knob, and turn it to take up any slack. Attempt to open the door in whichever direction it opens. If it is secure, check for pry marks and then move on to the next door or window.

Inspecting windows usually involves first checking for any broken glass, pry marks, or latent prints before touching any portion of the window to avoid contaminating possible evidence. Make sure there is actually glass in the window. It may have been completely removed by some careful burglar. Touch the glass to feel for any unusual heat inside the building, which may indicate the presence of a smoldering fire or a faulty heating system. Determine the way the window opens, then attempt to open it by applying reasonable pressure. If there is some evidence of entry, check the wall below the window for scuff marks and the sill for dust disturbance. Sometimes the window may prove to be one that was broken earlier but never repaired. At first it may appear to be a burglar's point of entry, but closer inspection may reveal a substantial quantity of dust and the presence of spider webs. These would have been destroyed if entry had been made recently.

If you have a partner, determine in advance exactly who is going to enter the premises. Officers should never enter from opposite sides of a building. If they plan to enter the building together, they should do so as a team. They should work closely together, constantly keeping an eye on each other. Either work side by side, or back to back. One may alternately take the lead with one officer behind the other, in a leapfrog pattern. The purpose of this method of coverage is to assure each partner that the other partner knows exactly who and where the other is in the darkened or deserted interior of the building. Without such a system it is possible that one officer could walk into the pointed barrel of another officer's revolver.

When you use a flashlight, hold it away from your body, preferably with your weaker hand. Most inexperienced shooters have a tendency to flinch to their right while shooting directly at a light source. When entering the building try to avoid making a silhouette of yourself by standing in the center of the doorway. At the moment you enter, consider turning off your flashlight. When you open the door, stand to one side, out of the line of fire, if the suspect should choose to attack you. Open the door by slamming it against the stop to make sure no one is hiding behind it. Step in quickly and immediately to one side, out of the door opening. Wait until your partner follows suit, then stand and listen for a few seconds to get accustomed to the new lighting conditions and the sounds inside of the building. There are some noises characteristic to the building that can be disconcerting if you are not accustomed to them. Refrigerators, freezers, electronic equipment, clocks, heaters, and air conditioning devices all sound differently in strange buildings and in tense situations.

Once you are inside and have determined the suspect is not in the room with you, make liberal use of your flashlight. When you turn on the electric lights as you go through, consider the advisability of turning them back off before entering the next room. Don't make yourself a silhouette target. Each time you enter another room, slam the door to the door stop or the wall. Again, there may be someone behind the door.

Make a thorough and methodical check of the premises. Determine whether a crime was committed and whether the suspect may still be inside. When people are hiding, they discover they can literally shrink into spaces they never thought possible. Look up; burglars sometimes hide on closet shelves or in spaces above closets. Attics, large drawers, incinerators, storage cabinets, almost anywhere is a potential hiding place. Once a room has been searched, clear the searched area by closing the door and placing something in front of it so the object would have to be moved to pass through.

If you find the suspect in the building, search him immediately and use your handcuffs. Do not assume the suspect is alone. Continue the search until the entire building is secure. Look for an accomplice outside and nearby.

Should it be there was no burglary, but the building was just left open, secure the premises and contact the owner. Tell the owner of your findings and request the owner make an appearance on the scene. Your company should have an emergency list of phone numbers of each of its clients.

INSPECTING RESIDENTIAL PROPERTY

People may absent their homes for a few hours or they may be gone for long periods of time on vacation, for illness, or other reasons. A crime committed against property can go undetected and unreported for all of that time. The greater the time between the occurrence of the crime and its discovery, the greater the chance the culprit will escape apprehension.

If residential property is contained within your patrol area, basic precautionary efforts and communication with local police are encouraged. Many people do not think about the need to communicate their comings and goings. They are lax by nature and then experience regret later.

Victims of burglaries are often their own worst enemy. They don't notify the letter carrier to stop delivery of the mail or halt newspaper delivery. They announce their travel plans to the entire neighborhood. They should make preparations to have the lawn mowed while they are away in addition to the other notifications mentioned. Inside houselights should be hooked to timers in the place where the residents normally spend a considerable amount of time. They should make their house look like they are home. If the patrol officer sees newspapers and mail accumulating and the grass uncut, the officer might talk to one of the neighbors to see whether the neighbor will take care of the homeowner's carelessness.

PROWLERS

If you know a neighborhood where a prowler has been reported, you have an advantage. With such knowledge it is possible to stop your vehicle a distance from where the prowler was reported seen and approach quietly on foot. The suspect may not be aware of your arrival. If you have a partner, he or she can take a position on a parallel street directly behind the house where you expect the prowler to be. Should the prowler go over the fence to escape, your partner might be there waiting.

Approach the suspect location quietly, driving your vehicle in the same manner as a local resident would do returning home. It may be to your advantage to stop a distance down the street, turn off the headlights, then roll along the curb with your lights out for the remaining distance. Use the emergency brake to avoid the brake lights going on.

As you approach, observe the people in the area. Are they making sneaky movements? Is their conduct suspicious? Anyone running, or sneaking around the spot a prowler was reported is surely suspect, but do not overlook the person who is acting perfectly normal as if nothing were wrong. That person could be the prowler.

If you are not able to locate the suspicious person quickly after your arrival, there is an excellent chance the person knows of your presence. This being the case, you can use plenty of light and conduct a thorough and systematic search of the area. Do not unnecessarily disturb the neighbors. Keep your physical activity to a minimum. Do not assume anyplace is too small to hide. Use the same search techniques as used in search of buildings. If you locate the suspect and are sure you have the right person, shine your flashlight in the suspect's eyes to keep the person at a disadvantage. Doing this makes it difficult for the suspect to make any moves against you.

A prowler in the early stages is a misdemeanant. There is no reason for a drawn weapon at this time. If the prowler makes an attempt to assault you or escape after you have placed him under arrest, *you may use reasonable force if absolutely needed to complete the apprehension or to defend yourself.*

CHAPTER 20 PROBLEMS

PROBLEM 1

A citizen complains to a roving patrol security officer that he has been receiving threatening phone calls for the past week. What action should the security officer take?

PROBLEM 2

Security officers are contracted to assist the local police in the maintenance of order in a public park during a Fourth of July celebration at which about 5,000 persons are in attendance.

A group of seven or eight people are drinking beer in the park in violation of the local ordinance. What should the security officers do?

PROBLEM 3

Security officer Holtz is patrolling his normal beat in an industrial complex when his dispatcher advises Holtz that a client has a problem and wishes to see him.

Holtz goes directly to the client's business establishment where he meets with the client. The client tells Holtz that his establishment has been closed for public business for an hour. He states that an irate citizen is still in the premises and refuses to leave. What should the security officer do?

PROBLEM 4

For the greatest degree of safety for an officer involved, a lone officer should not attempt to frisk more than

A. 1 person.
B. 2 persons.
C. 3 persons.
D. 4 persons.

PROBLEM 5

While on routine patrol you see a man, known to you to be a wanted person, walking along the side of the street. You stop your car and order the person to stop after first identifying yourself. The suspect takes off running with you in hot pursuit. Which of the following should you do?

A. At night, keep a steady beam of light on him.
B. Take the same path as the suspect.
C. Keep your revolver ready, in your hand.
D. Throw your handcuffs or nightstick at him, and try to stun him.

Cases Index

Y

Z

Author and Subject Index